LITHIUM-DRIFTED GERMANIUM DETECTORS

Their Fabrication and Use

An Annotated Bibliography

LITHIUM-DRIFTED GERMANIUM DETECTORS

Their Fabrication and Use

An Annotated Bibliography

Compiled by
Ina Calloway Brownridge

Associate Librarian
State University of New York at Binghamton
Binghamton, New York

IFI/PLENUM • NEW YORK • WASHINGTON • LONDON • 1972

Library of Congress Catalog Card Number 73-183565

ISBN-13: 978-1-4613-4600-5   e-ISBN-13: 978-1-4613-4598-5

DOI: 10.1007/ 978-1-4613-4598-5

© 1972 IFI/Plenum Data Corporation

Softcover reprint of the hardcover 1st edition 1972

A Subsidiary of Plenum Publishing Corporation

227 West 17th Street, New York, N.Y. 10011

United Kingdom edition published by Plenum Press, London

A Division of Plenum Publishing Company, Ltd.

Davis House (4th Floor), 8 Scrubs Lane, Harlesden, NW10 6SE,

London, England

PREFACE

A lithium-drifted germanium detector is a semiconductor device which operates at liquid nitrogen temperature, and is used for detection of nuclear radiation, mostly gamma ray. The detection occurs when the γ-ray undergoes an interaction in the intrinsic or I region of the semiconductor. The interaction results in the production of charge carriers which are swept out by an electric field. This is accomplished by reverse biasing the detector with approximately 100 v/mm of intrinsic material. The total amount of charge swept out is proportional to the energy dissipated in the intrinsic region. This may include the total energy of the photon, but generally somewhat less.

The Ge(Li) device is a semiconductor p-n device with a very large intrinsic region between the positive carrier region and the negative carrier region (P-I-N). The fabrication of this device consists of three major steps: the diffusion of the lithium into the p-type germanium to give an n-type surface region, the drifting process to obtain the intrinsic region as deeply as possible, and the surface preparation. There are numerous procedures for the various steps as well as criteria for material selection and the preparation of the materials.

Since the fabrication of the first lithium-drifted detectors in the early 1960's, a large volume of literature has been published and continues to be published on the subject. The literature in this guide represents publications from various research facilities in the United States and abroad -- national research laboratories, privately operated research facilities, as well as research facilities on college and university campuses, and it covers materials published to May 1971.

This guide is a compilation of references to works on the uses of the Ge(Li) device as a radiation detector, the fabrication of the detector and the associate electronics with its great varieties of uses. It consists of 790 entries to literature in English, French, German, Japanese, Russian, Spanish and several other languages, and it includes an author and subject index. The numbers in the indexes refer to entry number rather than page number.

Major bibliographic sources for the entries and abstracts are: Nuclear Science Abstracts, U. S. Government Research and Development Reports, Dissertation Abstracts, Scientific and Technical Aerospace Reports, as well as technical reports, and other pertinent primary publications. The entries are arranged alphabetically by author, except when there is no personal name, and title. Each entry includes author, if personal name or names are given, title, corporate author for technical reports and theses, and sources of information, e. g. citation for periodicals, serial number for technical reports and bibliographic data for books. Publications in languages other than English are indicated in parenthesis at the end of the citation.

I wish to thank Sol Raboy of the Physics Department of the State University of New York at Binghamton, and members of the Physics Department's staff for suggestions and technical assistance. Special thanks to James D. Brownridge for his many suggestions. In addition, I would like to thank Betty Lindsley, Marilyn McLeod and Charlotte Bunzey for their efforts in preparing the manuscript for publication.

The project would not have been feasible without the resources of the Library of the State University of New York at Binghamton and the cooperation of the library staff.

CONTENTS

ABBREVIATIONS

Acad. Repub. Pop. Rom. Stud.
   Cercet. Fiz. .... Academia Republicii Populare Romine,
                                  Institutul de Fizica Atomica si In-
                                  stitutul de Fizica, Studii si Cer-
                                  cetari de Fizica

Acta. Phys. Austr. .... Acta Physica Austriaca

Amer. J. Phys. .... American Journal of Physics

Anal. Chem. .... Analytical Chemistry

Ann. Acad. Sci. Fenn. Ser. A VI .. Annales Acadamiae Scientiarum Fen-
                                  nicae, Series A VI. Physica

Ann. Rev. Nucl. Sci. .... Annual Review of Nuclear Science

Appl. Phys. Lett. .... Applied Physics Letters

Ark. Fys. .... Arkiv fuer Fysik, Utgivet au Kun-
                                  gliga Svenska Vetenskapsakademien

At. Energ. (USSR) .... Atomnaya Energiya (U. S. S. R.)

At. Energy .... Atomic Energy

At. Energy Rev. .... Atomic Energy Review

Atomics .... Atomics

Atomkernenergie .... Atomkernenergie (formerly Atomener-
                                  gie)

Atoomenergie Haar Toepass. .... Atoomenergie en Haar Toepassingen
                                  (formerly RCN Bulletin)

Brit. J. Radiol. .... British Journal Radiology

Bunseki Kiki .... Bunseki Kiki (Analysis and Instru-
                                  ment)

CERN Cour. .... CERN Courier

Can. Chem. Process. .... Canadian Chemical Processing

Can. J. Phys. .... Canadian Journal of Physics

Contemp. Phys. .... Contemporary Physics

Dokl. Akad. Nauk. SSSR .... Doklady Akademii Nauk SSSR

Electron. Lett. .... Electronics Letters

Fiz. Tekh. Poluprov. .... Fizika i Tekhnika Poluprovodnikov
                                  (Engl. Trans. Soviet Physics-Semi-
                                  conductors)

Fizika (Zagreb) .... Fizika (Zagreb)

Health Physics .... Health Physics

Helv. Phys. Acta .... Helvetica Physica Acta

IEEE Proceedings .... Proceedings of the Institute of
                                  Radio Engineers

IEEE Trans. Nucl. Sci. .... IEEE Transactions on Nuclear Science
                                  (formerly IRE (Institute of Radio
                                  Engineers) Transactions on Nuclear
                                  Science)

Ind. Atom. .... Industries Atomiques

Ind. Chim. Belge .... Industrie Chimique Belge

Indian J. Phys. .... Indian Journal of Physics and Pro-
                                  ceedings of the Indian Association
                                  for the Cultivation of Science

Instrum. and Exper. Tech. .... Instruments and Experimental Tech-
                                  niques

IRE Trans. on Nucl. Sci. .... SEE IEEE Transactions on Nuclear
                                  Science

Izv. Akad. Nauk SSSR. Ser. Fiz. .. Izvestiya Akademiia Nauk SSSR,
                                  Seriya Fizicheskaya

Jad. Energ. .... Jaderna Energie

J. Appl. Phys. .... Journal of Applied Physics (U. S.)

J. Appl. Phys., Japan .... Journal Applied Physics, Japan

J. Belge Radiol. .... Journal Belge de Radiologie

J. Natur. Sci. Math. .... Journal of Natural Sciences and
                                   Mathematics

J. Nucl. Med. .... Journal of Nuclear Medicine

J. Nucl. Sc. Seoul) .... Journal of Nuclear Sciences (Seoul)

J. Nucl. Sci. Technol. (Tokyo) ... Journal of Nuclear Science and
                                   Technology (Tokyo)

Kanagawa-Ken Kogyo Shikenjo Kenkyu
   Hokoku .... Kanagawa-Ken Kogyo Shikenjo Kenkyu
                                   Hokoku (Bulletin of the Industrial
                                   Research Institute of Kanagawa
                                   Prefecture (Japan))

Kerntechnik .... Kerntechnik, The Journal for Nu-
                                   clear Engineers and Scientists

Latv. PSR Zinat. Akad. Vestis, Fiz.
   Teh. Zinat. Ser. .... Latvijas PSR Zinatnu Akademijas
                                   Vestis, Fizikas un Tehnisko Zinatnu
                                   Serija

Magy. Radiol. .... Magyar Radiologia

Nature .... Nature

Naturwiss. Rundsch. .... Naturwissenschaftliche Rundschau

Ned. Tijdschr. Natuurk. .... Nederlands Tijdschrift voor Natuur-
                                   kunde

Nucl. Appl. .... Nuclear Application

Nucl. Eng. .... Nuclear Engineering Incorporating
                                   Nuclear Power (England)

Nucl. Instrum. Methods .... Nuclear Instruments and Methods

Nucl. Phys. .... Nuclear Physics (Netherlands)

Nucl. Sci. Appl. .... Nuclear Science and Applications

Nucleus (Paris) .... Nucleus. La Revue Scientifique a
                                   l'Age Atomique (Paris)

Nukleonik .... Nukleonik

Nukleonika .... Nukleonika (Poland)

Ohio J. Sci. .... Ohio Journal of Science

Onde Elect. .... Onde Electrique, L'

Oyo Butsuri .... Oyo Butsuri (Journal of Applied
                               Physics (Japan))

Phys. Lett. .... Physics Letters

Phys. Rev. .... Physics Review

Phys. Today .... Physics Today

Prib. Tekh. Eksp. .... Pribory i Tekhnika Eksperimenta

Proc. IEEE (Inst. Elec. Electron.,
    Eng.) .... Proceedings of the IEEE (Institute
                                of Electrical and Electronic Engi-
                                neers, formerly Proceedings of the
                                Institute of Radio Engineers)

Rc. Accad. Naz. Lincei (Italy) ... Accademia Nazionale Dei Lincei
                                   (Italy) Rendiconti

Radiobiologiya .... Radiobiologiya

Radiochim. Act .... Radiochimica Acta

Rep. Progr. Phys. .... Reports on Progress in Physics

Rev. Phys. Appl. .... Revue de Physique Appliquee (Supple-
                               ment au Journal de Physique)

Rev. Roum. Phys. .... Revue Roumaine de Physique

Rev. Sci. Instrum. .... Review of Scientific Instruments

Sci. Rep. Yokohama Nat. Univ.
    (Japan) .... Science Reports of the Yokohama
                                National University

Science .... Science

Siemens Rev. .... Siemens Reviews

Sov. Phys.-Semicond. .... Soviet Physics-Semiconductors

Suomen Kem. .... Soumen Kemistilehti

Talanta .... Talanta

Toshiba Rebyu .... Toshiba Rebyu (Toshiba Review)

Trans. Amer. Nucl. Soc. .... Transactions of the American Nuclear
Society

Vestsi Akad. Navuk BSSR, Ser. Fiz-
Mat. Navuk .... Vestsi Akademii Navuk BSSR, Seriyya
Fizika-Matematychnykh Navuk

Z. Anal. Chem. .... Zeitschrift fuer Analytische Chemie,
Fresenius

Z. Angew. Math. Phys. .... Zeitschrift fuer Angewandte Mathe-
matik und Physik

Z. Phys. .... Zeitschrift fuer Physik

Zh. Tekh. Fiz. .... Zhurnal Tekhnicheskoi Fiziki

BIBLIOGRAPHY

LITHIUM-DRIFTED
GERMANIUM DETECTORS

1. Adams, F. FABRICATION, PROPERTIES AND APPLICATIONS OF GE(LI)
   GAMMA DETECTORS. At. Energy Rev., 5 (1967), 31-92.

   Gamma-ray spectrometry has progressed rapidly as a result
   of the development of lithium-drifted germanium detectors. In
   both pure nuclear research and in applied spectrometry, the
   excellent resolution and the good detection efficiency of these
   devices is now fully recognized. The fabrication of detectors
   of different types, such as planar and coaxial, taking into account the practical difficulties such as lithium precipitation,
   drift failure, and surface leakage current is described. The
   instrumentation including preamplifier, amplifier analyzing
   equipment, and the problems associated with very high resolution
   spectrometry are briefly reviewed. The characteristics of different types of detectors, such as energy and time resolution
   and detection efficiency are mentioned. A number of applications
   in analytical and radiochemistry are discussed. These include
   activation analysis, x-ray fluorescence analysis, and fission
   product studies.

2. Adams, F. GERMANIUM DETECTORS IN CHEMICAL ANALYSIS. Paper 3
   of "The Proceedings of the Meeting on the Application of
   Ge(Li) Detectors in Science, Technology, Medicine and Industry, Brussels, Oct. 20, 1967." 15p.
   BLG-425; CONF-671078

   The uses of lithium-drifted germanium detectors in activation analysis and x-ray fluorescence analysis are summarized.

3. Adams, F. A MOUNTING PROCEDURE FOR GE(LI) DETECTORS. Nucl.
   Instrum. Methods, 48 (1967), 338-340.

4. Adams, F. PROPERTIES OF GERMANIUM DETECTORS AND THEIR USES FOR
   GAMMA SPECTROMETRY. J. Belge. Radiol., 51 (1968), 40-8.
   (In French)

The properties of germanium detectors, their manufacture
and uses for gamma spectrometry are described. At the present
time, there are germanium detectors available that are equal
to the classical scintillation detectors for the measurement by
gamma spectrometry.

5. Adda, L. P.; Benson, K. E.; deWit, R. C., and McKenzie, J. M.
   GROWTH OF GERMANIUM FOR LITHIUM DRIFT DETECTORS. IEEE
   Trans. Nucl. Sci., NS-15, no. 3 (June 1968), 347-351.

   The germanium for lithium drift germanium detectors should
   have the following properties:

   1. High lithium drift rate
   2. Low reverse current during the drifting
      process- particularly for high temperature
      drifting
   3. Minimal lithium precipitation during the drifting
   4. Capability of fabrication into detectors exhibiting:
      (a) Low noise reverse currents at high bias
          voltages
      (b) A minimum of carrier trapping

6. ADVANCES IN SCINTILLATION AND SEMICONDUCTOR RADIATION DETECTOR.
   Atomics, 18 (July-Aug. 1965), 11-19.

   Progress in methods of detecting and measuring ionizing
   radiation using lithium-drifted germanium diodes is reviewed
   along with improvements to photomultipliers. Scintillation
   detector development is described.

7. Ahmad, A. A. Z. and Ahmed, N. M.   THEORETICAL CALCULATION OF THE
   LOW ENERGY EFFICIENCIES OF CYLINDRICAL GE(LI) DETECTOR.
   Pakistan Inst. of Nuclear Science and Technology, Islamabad.
   May 1968. 20p.   PINSTECH/PHY-16

   Efficiencies for a cylindrical Ge(Li) detector for gamma
   rays were calculated theoretically. An IBM 1620 computer was
   used for obtaining numerical values. Efficiencies were found
   for different sizes of the detector and also for different
   values of source to crystal distance and energy.

8. Ahmed, Nighat Masud; Abul Faiz Mohammed; Najam, Mohammad Ramzan;
   Khan, Zafarullah, and Khan, Hameed Ahmed. GE(LI) DETECTOR

USED IN A PAIR SPECTROMETER CONFIGURATION. Pakistan Inst. of Nuclear Science and Technology, Islamabad. May 1968. 11p.   PINSTECH/PHY-13

A three crystal pair-spectrometer with a Ge(Li) detector as the central crystal has been set up. It has been used to record the pair spectrum of ^{72}Ga. The results obtained and the relative efficiency calibration of the spectrometer are discussed.

9. Ahmed, Nighat Masud; Khan, H. A.; Ishaq, A. F. M.; Ahmad, A. A. Z., and Najam, R. THE LOW LYING ENERGY LEVELS IN GE 72. Pakistan Inst. of Nuclear Science and Technology, Islamabad. May 1968. 29p.   PINSTECH/PHY-14

The decay scheme of ^{72}Ge was studied using Ge(Li) and Na(Tl) detectors. The results of the singles and pair-spectrometer measurements reveal 33 gamma lines with energies lying between 286 and 2837 keV. These were fitted into a decay scheme on the basis of the results of gamma-gamma coincidence experiments.

10. Alexander, T. K. THE IMPACT OF GERMANIUM DETECTORS AT CHALK RIVER. pp. 780-783 of "Semiconductor Nuclear-Particle Detectors and Circuits, Proceedings." Brown, W. L., ed. Washington, D. C., National Academy of Sciences, 1969.

11. Alexander, T. K.; Broude, C.; Haeusser, O., and Sharpey-Schafer, J. F. PAIR AND ESCAPE SUPPRESSED SPECTROMETER USING GE(LI) AND NA(TL) DETECTORS DESIGNED FOR ACCELERATOR EXPERIMENTS. Nucl. Instrum. Methods, 65 (Nov. 1, 1968), 169-72.

A high resolution gamma-ray spectrometer consisting of a 40 cm^3 Ge(Li) detector located at the common center of two semi-annular Na(Tl) detectors has been assembled for use in particle accelerator experiments.

12. Alkhozor, G. D.; Komar, A. P., and Vorobev, A. A. IONIZATION FLUCTUATIONS AND RESOLUTION OF IONIZATION CHAMBERS AND SEMI-CONDUCTOR DETECTORS. Nucl. Instrum. Methods, 48 (1967), 1-12.

The Fano factor F determining the variance of ionization fluctuations is calculated by Fano's approximate method for a variety of gases and for Si and Ge.

13. Allen, B. J.; Upex, G. D., and Trimble, G. D. THE ANALYSIS OF
    GE(LI) GAMMA RAY SPECTRA WITH AN ON-LINE COMPUTER. Atomic
    Energy Commission Research Establishment, Lucas Heights,
    (Australia). June 1968. 19p.   AAEC/TM-458

    The application of a small on-line computer to data-taking
    and analysis of Ge(Li) gamma ray spectra is described. Both
    symbolic and FORTRAN programs are included and the merits of
    on-line analysis discussed.

14. Allen, B. J. HIGH ENERGY RESPONSE FUNCTIONS OF A GERMANIUM
    GAMMA RAY SPECTROMETER. Nucl. Sci. Appl., 3 (Oct. 1967),
    74-9.

    Response functions for high energy gamma rays of a coaxial
    germanium diode have been obtained and analyzed in terms of
    photon absorption mechanisms. Line shapes and peak efficien-
    cies are given to 17.6 MeV. Double escape peak efficiencies
    are related to spectra obtained when the detector is operated
    as a pair spectrometer.

15. Allen, B. J. THE HIGH RESOLUTION OF GAMMA RAYS FROM KEV NEUTRON
    CAPTURE. Atomic Energy Comm. Research Establishment, Lucas
    Heights, (Australia). July 1968. 31p.   AAEC/TM-462

    A Ge(Li) detector and time-of-flight system are used with
    an on-line computer to measure gamma rays emitted after keV
    neutron capture. The low efficiency of the Ge(Li) detector
    necessitates careful optimization of experimental parameters.
    Conflicting requirements of yield rate and timing resolution
    are discussed in terms of accelerator performance, target geom-
    etry and data-taking methods. The selection of suitable capture
    targets is considered.

16. Allen, B. J.; Bird, J. R., and Engstrom, S. RESPONSE FUNCTIONS
    OF A GERMANIUM-SODIUM IODIDE DETECTOR SYSTEM. Nucl. Instrum.
    Methods, 53 (1967), 61-70.

    A gamma ray spectrometer is described which uses a lithium
    drifted germanium diode or a sodium iodide crystal as a central
    detector in conjunction with an annular segmented sodium iodide
    assembly. The system can operate as a total absorption, anti-

coincidence, or pair spectrometer and individual detectors may be used separately. Thus, the requirements of high resolution or high efficiency gamma ray spectroscopy can be met by suitable choice of mode of operation. The various modes of operation are compared and typical results given to illustrate their performance at a variety of gamma ray energies.

17. Allkofer, O. C.; Fox, J. M., and Grupen, C. A TRIGGER-CIRCUIT FOR BETA-RAY SPECTROMETERS WITH SEMICONDUCTOR DETECTORS IN STACK-ASSEMBLIES. Nucl. Instrum. Methods, 47 (1967), 342-344.

When taking beta-ray spectra by semiconductor detector techniques, it is an advantage to stack several detectors in series. The resolution of a single detector determines the optimum resolution of the stack assembly. However, such a system involves also the possibility of spurious counts. These edge effect cause an additional deterioration of the resolution. To take best out of the resolution of the detectors a trigger-circuit was developed, to exclude the spurious pulses from entering the pulse-height analyzer. A coincidence-anti-coincidence-circuit discrimination against the unwanted pulses. The pulses so selected are fed to the imput gate of the PHA.

18. Ammerlaan, C. A. J.; Rumphorst, R. F., and ch Koerts, R. F. PARTICLE IDENTIFICATION BY PULSE SHAPE DISCRIMINATION IN THE P-I-N SEMICONDUCTOR. Nucl. Instrum. Methods, 22 (Apr. 1963), 189-200.

When an ionizing particle enters a semiconductor radiation detector, a voltage is produced. The final pulse height is determined by the particle energy, while the range of the particle affects the shape of the pulse. The voltage pulse therefore contains information about the particle type. In the present paper the pulse height at a fixed time during the pulse rise time, is used as the range dependent quantity. A two-dimensional display of this quantity versus the energy of the particle showed that separation could be obtained between deutrons and α particles with energies in this range from 8 to 26 MeV.

19. Ammerlaan, C. A. J. and Mulder, K. THE PREPARATION OF LITHIUM-DRIFTED SEMICONDUCTOR NUCLEAR PARTICLE DETECTORS. Nucl. Instrum. Methods, 21 (1963), 97-100.

A method is described for the preparation of p-i-n structure

semiconductor nuclear particle detectors using the process of
drift. The procedure is described in detail and much attention
is given to the measurements of parameters which control various steps in the procedure. The reproducibility, obtained by
the method, turned out to be very high. The properties of the
detectors are described briefly.

20. Andrieux, Claude and Dumesmil, Patrice. GAMMA-RAY SPECTRO-
    METRY IN DRILLINGS USING GE(LI) SEMICONDUCTOR PROBES.
    Commissariat a l'Energie Atomique, Saclay, France. Centre
    d'Etudes Nucleaires. July 1, 1970, 26p. (In French)
                        CEA/Conf-1694; Conf.-701030-1

    From Meeting on the Use of Radioelements and Nuclear Methods
in Geophysics for the Prospection and Exploitation of the Minerals, Nancy, France.

    A Ge(Li) semiconductor well logging probe possessing ex-
cellent energy resolution characteristics is described. The
method used to determine the uranium content of rocks is discussed. In order to demonstrate the superiority of this method
of analysis, comparisons were made between the spectra obtained
from a sample containing 0.5% uranium using a scintillographic
probe, and the Ge(Li) semiconductor probe.

21. Anders, O. U. EXPERIENCES WITH THE GE(LI) DETECTOR FOR HIGH-
    RESOLUTION GAMMA RAY SPECTROMETRY AND A PRACTICAL APPROACH
    TO THE PULSE PILE-UP PROBLEM. Nucl. Instrum. Methods, 68
    (Feb. 15, 1969), 205-8.

    Ageing of a Ge(Li) detector results in a decrease of the
signal from the detector at a rate of 1% per year. Pulse pile-
up at high counting rates distorts all peaks in a typical spectrum in the same way depending on the counting rate and the o-
verall shape of the spectrum. Pulser timing corrects for both
analyzer dead time and pulse pile-up distortion and thus gives
the correct timing of the counting interval if peak areas are
to be used for quantitative work. Leaky discriminators are
proposed for multichannel analyzers to permit effective usage
of two-pulser spectrum stabilization.

22. Andersen, B. V.; Bramson, P. E., and Unruh, C. M. COMPARISON
    OF GERMANIUM AND SODIUM IODIDE IN VIVO MEASUREMENT SYSTEMS.
    Battelle-Northwest, Richland, Wash. Pacific Northwest Lab.,
    1970. 19 p.   CONF-701112-1; BNWL-SA-3397

From Symposium on New Developments in Physical and Biological Radiation Detectors, Vienna, Austria.

Lithium-drifted germanium Ge(Li) and lithium-drifted silicon Si(Li) counting systems for in vivo measurements are compared with conventional scintillator detector systems in similar configurations. Measurements of plutonium and americium in lungs, other organs, and wounds using coaxial and planar drift detectors are presented. A proposed large area planar Ge(Li) lung counter system is compared to 2 and 4 crystal sodium iodide counters (130 cm^2 by 9 mm thick) currently used for uranium and plutonium lung measurements. A large Ge(Li) detector system being employed at Battelle-Northwest Laboratory to make whole body measurements of radio nuclide depositions in humans consists of four coaxial detectors each 40 cm^3 in volume for a total of 160 cm^3. The individual detectors are enclosed in separate cryostats but mounted in a common 30 liter liquid nitrogen dewar of the "chicken feeder" design. The system is compared to the standard 23 cm diameter by 10 cm thick sodium iodide scintillator in the standard chair position. (auth.)

23. Andersen, B. V.; Bramson, P. E., and Rising, F. I. IN VIVO COUNTING WITH GE(LI). Battelle-Northwest, Richland, Wash. Pacific Northwest Lab. May 27, 1970. 9p.
   BNWL-SA-3069; CONF-700609-2

This report describes a whole-body counter using four Ge(Li) detectors. The results are compared to those of a sodium iodide scintillation detector.

24. Anderson, William R. THE DETERMINATION OF A PARTIAL DECAY SCHEME OF Dy [160]. State Univ. of New York at Binghamton. 1971. 84p. Thesis

The determination of a particle decay scheme of Dy[160] was made using data collected at the nuclear spectroscopy facility at the State University of New York at Binghamton. The gamma radiation associated with the beta decay of Tb[160] to levels in Dy[160] was studied with Ge(Li) spectrometers. The data was recorded by an IBM 1800 computer. This was the initial experiment performed at the facility, and the results were compared to previously reported results.

25. Anicin, I. V. and Bikit, I. S. A NOTE ON THE FINITE SOLID-ANGLE CORRECTIONS FOR GE(LI) DETECTORS. Nucl. Instrum. Methods, 87 (1970), 145.

26. Antman, S. O. W.; Landis, D. A., and Pehl, R. H. MEASUREMENTS
    OF THE FANO FACTOR AND THE ENERGY PER HOLE-ELECTRON PAIR IN
    GERMANIUM. Nucl. Instrum. Methods, 40 (1966), 272-276.

    A Fano factor of 0.30±0.03 has been measured for germanium
    with a lithium-drifted semiconductor detector used for measur-
    ing gamma-ray energies ranging from 100-2800 keV. A value of
    2.98±0.01 eV for the average energy per hole-electron pair at
    77°K was measured for gamma ray peaks in the energy region from
    100-1400 keV.

27. Antonov, A. S. and Yuskeselieva, L. G. PRIMENENIE ELEKTRO-FOTO-
    GRAFII DLYA IZMERENIYA CHUVSTVITEL' NOL OBLASTI GERMANIEVYKH
    P-I-N DETEKTOROV. (APPLICATION OF ELECTROPHOTOGRAPHY FOR
    MEASUREMENTS OF SENSITIVE REGION OF GERMANIUM P-I-N DETEC-
    TORS.) Jt. Institute for Nucl. Research, Dubna (USSR) Lab.
    of Nucl. Problems. 1965. 10p.   JINR-P-2487

    The width of the sensitive region in germanium p-i-n detec-
    tors is determined by developing the p-i-n junction with dry
    and liquid electrophotographic developers.

28. Armantrout, Guy A. AMBIENT STORAGE EFFECTS AND MOUNTING PRO-
    BLEMS OF VERY LARGE VOLUME GE(LI) DETECTORS. California
    University. Lawrence Radiation Lab. Oct. 8, 1965. 34p.
                                                UCRL-14263

        From IEEE Nucl. Sc. Symposium, San Francisco.

    A study was undertaken of the surface-setting procedure
    before mounting and the effect of diode storage for varying
    periods of time at the ambient temperatures was investigated.
    The diode surface potential, diode capacitance, and diode V-1
    characteristics were taken as a function of different surface
    treatments before mounting. The results were correlated using
    an inversion layer model for the diode leakage currents less
    than 0.5nA at 3 KV were obtained. The compensated region of
    Ge(Li) D diodes is slightly supersaturated with lithium at room
    temperature. These diodes tend to revert back to the original
    p-type material because of lithium precipitation. The rate of
    precipitation was found for several diodes and was found to be
    a function of the effective lithium mobility.

29. Armantrout, Guy A. CORRELATION BETWEEN LITHIUM DRIFT MOBILITY
    AND MINORITY-CARRIER DRIFT MOBILITY IN GERMANIUM. IEEE
    Trans. on Nucl. Sci., NS-13, no. 3 (June 1966), 370-372.

    The minority-carrier drift mobility and non-driftable ger-
    manium was measured in an effort to determine the presence of
    an impurity which reduced the lithium drift mobility but did not
    affect the resistivity and lifetime of some of the materials
    used to make lithium-drift detectors. The measurements were
    made at 77°K where the mobility is limited by impurity scatter-
    ing rather than lattice scattering.

    The mobility was measured as a function of temperature for
    five samples of Ge which had varying lithium drift mobilities.
    Good correlation was found between the driftability of the ma-
    terial and the minority-carrier drift mobility at 77°K. Esti-
    mated impurity levels in the range of 10^{14}-10^{15}/cm^3 apparently
    reduce the effective lithium mobility, while impurity concen-
    trations greater than 5×10^{15} make the Ge unsuitable for making
    lithium drift detectors.

30. Armantrout, Guy A. DEFECT STUDY AND IDENTIFICATION IN GE(LI)
    P-N JUNCTION RADIATION DETECTORS. California University,
    Lawrence Radiation Lab. September 3, 1969, 33p.
                                        CONF-690821;UCRL-71679

    From Electronic Materials Conference, Boston, Mass.

    Ge(Li) detectors, which are very high resolution gamma ray
    spectrometers, are extremely sensitive to defects in the junc-
    tion of the detector. These defects act as charge-trapping
    centers and seriously degrade the detector performance. Several
    techniques, which include the detector monochromatic infrared
    response, reverse current characteristics, and performance analy-
    sis have been used to study these defects. The results indicate
    that degradation of the detector spectrum is due both to detect-
    or inhomogeneities and traps. The main inhomogeneity is caused
    by a localized variation in the precipitation rate of lithium,
    which results in a weak detector collection fields and heavier
    localized trapping.

31. Armantrout, Guy A. EVALUATION OF TRAPPING AND TAILING EFFECTS
    IN GE(LI) DETECTORS. California University, Livermore,
    Lawrence Radiation Lab. Jan. 22, 1969. 11p.
                                        UCRL-71508

    Day's model of trapping effects in detectors has been sim-
    plified and applied to the case of Ge(Li) detectors. With this

model it is possible to easily determine τ_t for any given detector. τ_t, the mean time before trapping of a carrier, is related to the trap density and trap capture cross section and is a useful parameter for characterizing the trapping properties of Ge. A number of ingots have been evaluated. In general, horizontally grown ingots closely follow the expected behavior of the model, and values of τ_t = 10 to 30 μsec have been found. However, vertically grown ingots, which have larger values of τ_t (>50 μsec), appear to be subject to thermal defects which ultimately result in nonlinear collection fields. The nonlinear fields cause large deviations from the response predicted by the simple model and are the cause of much of the tailing seen in vertically pulled material.

32. Armantrout, Guy A. A FABRICATION TECHNIQUE FOR SIGNIFICANTLY REDUCING THE CAPACITANCE OF LARGE-VOLUME GE(LI) DETECTORS. California Univeristy, Livermore, Lawrence Radiation Lab. Feb. 10, 1966. 17p.   UCRL-14561

33. Armantrout, Guy A. A HIGHLY SENSITIVE PHOTORESPONSE TECHNIQUE FOR DETERMINING IMPURITY AND DEFECT ENERGIES AND CONCENTRATIONS IN GERMANIUM. California University, Livermore, Lawrence Radiation Lab. July 17, 1969. 22p.
                                         CONF-690808; UCRL-71625

    From 3rd International Conference on Photoconductivity, Stanford, California.

34. Armantrout, Guy A. INFRARED EVALUATION TECHNIQUES FOR GE(LI) DETECTORS. IEEE Trans. Nucl. Sci., NS-17, no. 1 (Feb. 1970), 16-23.   Also report CONF-691017-Pt. 4; UCRL-71822

    From 16th Nuclear Science Symposium, San Francisco, Calif.

    The types of Ge(Li) detector spectrum degradation due to trapping effects are first considered, and a model is presented which explains this degradation in terms of specific phenomena. Then two detector evaluation techniques which utilize infrared light for studying these phenomena are outlined and the significant results obtained using these techniques are presented...

35. Armantrout, Guy A. and Okudo, D. PREAMPLIFIER NOISE EFFECTS ON SPECTROMETER ENERGY RESOLUTION. IEEE Trans. Nucl. Sci.,

NS-18, No. 1 (Feb. 1971), 170-174.

36. Armantrout, G. A. and Thompson, H. W., Jr. SPECTRUM DEGRADATION
    EFFECTS IN GE(LI) DETECTORS. California University, Liver-
    more, Lawrence Radiation Lab. March 1970. 24p.
                            CONF-700301-6; UCRL-72109

    From the 12th Scintillation and Semiconductor Counter Sym-
    posium, Washington, D. C.

    The mechanisms responsible for the degradation of the per-
    formance of Ge(Li) detectors (which include the effects of uni-
    form trapping, non-uniform trap distribution, and locally weak
    collection fields) are considered and simulated spectra have
    been computed. Experimental measurements have been made to
    identify these sources of degradation. The important results
    include a measurement of the effective capture cross sections of
    the most important traps and a determination of the main sources
    of non-uniform trap distribution and weak collection fields in
    Ge(Li) detectors.

37. Armantrout, Guy A. TRAPPING AND TAILING EFFECTS IN GE(LI) DE-
    TECTORS. Lafayette, Ind., Purdue Univ. 1969.   217p.
    Thesis.

    Ge(Li) detectors, which are very high resolution gamma-ray
    spectrometers, are very sensitive to any crystal impurities or
    defects which result in charge carrier trapping in the detector.
    The detector performance degradation results from the statisti-
    cal variation in the quantity of charge produced and from in-
    homogeneities in the original material. A more exact model was
    needed to determine the important trap carrier capture parameters.
    Measurements were developed or adapted to study specific aspects
    of the detector performance and to measure the pertinent trap
    parameters. These measurements include: infrared absorption of
    the initial germanium, response of the detector to monochromatic
    infrared light, infrared transmission television, detector re-
    verse current versus temperature measurements, photoconductive
    decay time, electrolytic etching and copper staining decoration
    of the crystal, and direct detector performance evaluation.

38. Armantrout, Guy A. TRAPPING EFFECTS IN GE(LI) DETECTORS: AN
    INTERIM PROGRESS REPORT. California University, Livermore,
    Lawrence Radiation Lab. Jan. 20, 1969. 40p.
                            CONF-690104-1; UCRL-71507

From American Society for Testing Materials Meeting, Gaithersburg, Md.
A discussion of the causes and nature of the trapping and
tailing problems associated with Ge(Li) detectors.

39. Armantrout, Guy A. U-JUNCTION GE(LI) DRIFT DETECTORS. IEEE
    Trans. Nucl. Sci., NS-14, no. 1-2 (1967), 503.
                              Also report CONF-661020-6; UCRL-14926

    Presented at the 13th Nucl. Sci. Sym., Boston, Oct. 19-21,
    1966.
    A U-Junction drifting configuration was used which permits
    the fabrication of large volume, low-capacitance detectors in
    much less time than previously required.

40. Armantrout, Guy A. and Camp, D. C. WINDOWLESS HIGH-RESOLUTION
    LID-GERMANIUM DETECTOR. Phys. Letters, 17 (July 15, 1965),
    279-280.   Also report UCRL-12333

    A description is given of a windowless lithium-drifted ger-
    manium diode, which is suitable for beta-ray spectroscopy. The
    evaluation of the detector resolution was made using a ^{57}Co and
    a ^{60}Co source is discussed.

41. Arnell, S. E.; Hardell, R.; Hasselgren, A.; Jonsson, L., and
    Skeppstedt, O. THERMAL NEUTRON CAPTURE GAMMAS MEASURED
    WITH GE(LI)-SPECTROMETERS AND INTERNAL REACTOR TARGETS.
    Nucl. Instrum. Methods, 54 (1967), 165-180.
                                        Also report AFCRL-67-0104

42. Arsent'ev, I. N.; Dneprovskii, I. S.; Popeko, L. A., and
    Samoilov, P. S. GE(LI) SPECTROMETER WITH A THERMOELECTRIC
    REFRIGERATING SYSTEM. At. Energ. (USSR), 28 (Feb. 1970),
    165-8. (In Russian)

43. Asikainen, M. and Blomqvist, L. MEASUREMENTS OF AIRBORNE RADIO-
    NUCLIDES IN FINLAND WITH A HIGH-VOLUME AIR SAMPLER AND GE(LI)
    SPECTROSCOPY. Institute of Radiation Physics, Helsinki,
    Finland, Feb. 1970. 40p.   SFL-A-15

    The use of a high-volume air sampler and Ge(Li) gamma spec-
    troscopy allows very sensitive measurement of several radionu-
    clides in air and makes possible the detection of debris from
    the nuclear explosions still being conducted in the atmosphere.

The methods used in sampling and measurements are described, and the results obtained in 1968-1969 are presented.

44. Atomic Energy of Canada, Ltd. Chalk River Nuclear Labs. PHYSICS DIVISION PROGRESS REPORT. April-June 1968. 1968. 88p.
                                                        AECL-3157

45. Atomic Energy of Canada, Ltd. Chalk River Nuclear Labs. PHYSICS DIVISION PROGRESS REPORT. Oct. 1, 1965-Dec. 31, 1965. 58p.
                                                        PR-P-68; AECL-2612

The direct γ spectrum from the reaction ^{25}Mg $(\alpha,n\gamma)^{28}$Si was measured as a function of the α-particle energy and the angle of the emitted γ-rays using a 25cm^3 Li-drifted Ge γ detector.

46. Atomic Energy of Canada, Ltd. Chalk River Nuclear Labs. PHYSICS DIVISION PROGRESS REPORT. Oct. 1, 1967-Dec. 31, 1967. 81p.
                                                        PR-P-76; AECL-3009

47. Atomic Energy of Canada, Ltd. PHYSICS DIVISION PROGRESS REPORT, Jan. 1-Mar. 31, 1968. 85p.
                                                        AECL-3108

The energy levels of several medium-weight and heavy nuclei were investigated using charged particles induced reactions and scattering as well as beta decay. Photofission and fission induced by direct reactions were also studied. Techniques for preparing germanium were investigated, as were pulse shapes in Ge(Li) detectors. A Ge(Li) double diode (n-i-p-i-n) was constructed. Neutron scattering from vanadium dioxide was studied, as was neutron inelastic scattering from liquid helium. The investigation of spin-waves in the conical magnetic phase of erbium was continued. Methods for making shell-model calculations were developed, as were methods for analyzing accelerator fields and beam transport systems.

48. Aubin, G; Barrette, J.; Lamoureux, G.; Monaro, S. CALCULATED RELATIVE EFFICIENCY FOR COAXIAL AND PLANAR GE(LI) DETECTORS. Nucl. Instrum. Methods, 76 (1969), 85-92.

The relative efficiency to monoenergetic gamma rays of coaxial and planar Ge(Li) detectors was calculated with the aid of a Monte Carlo computer program. Considering a point source, placed on the extended axis of the detector, at a fixed distance, radiations of different energies are generated within the solid

angle subtended by the crystal. The gamma-rays impinging on the
detector are followed until a fraction whatsoever of their energy escapes from the active region of the crystal.

49. Aubin, G.; Barrette, M.; Barrette, J., and Monaro, S. PRECI-
    SION MEASUREMENTS OF GAMMA RAY INTENSITIES AND ENERGIES IN
    THE DECAY OF ^{125}g, $^{154}Eu^{56}$, ^{110m}Co, Ag AND ^{125}Sb. Nucl.
    Instrum. Methods, 76 (1969), 93-9.

    With the aid of a computerized program the relative inten-
    sities of the prominent transitions present in the decay of
    ^{152}g, ^{154}Eu, ^{56}Co, ^{110m}Ag, and ^{110}Sb were calculated with high
    precision. Each of these sources could be employed both for
    energy calibration measurements and for linearity calibration
    of multichannel analyzers. The measurements were performed with
    two Ge(Li) coaxial detectors with active volume of $5.3 cm^3$ and
    $30 cm^3$, respectively. It is shown that with the aid of these
    radioactive sources, one can construct a very precise efficiency
    curve for any other Ge(Li) detector from 100 keV to 3500 keV
    and reproduce to within 2 to 3% the measured gamma ray relative
    intensities in ^{22}Na, ^{207}Bi, ^{133}Ba, and ThC". For these latter
    measurements a third Ge(Li) coaxial detector, with active vol-
    ume of $32 cm^3$, was employed.

50. Auble, R. L.; Beery, D. B.; Berzins, G.; Beyer, L. M.; Ether-
    ton, R. C., and Kelly, W. H. COINCIDENCE-ANTICOINCIDENCE
    GAMMA RAY SPECTROSCOPY WITH A NAI(TL) SPLIT ANNULUS AND A
    GE(LI) DETECTOR. Nucl. Instrum. Methods, 51 (1967), 61-71.

    A versatile gamma-ray spectrometer system consisting of a
    Ge(Li) detector surrounded by a large (20.3cm x 20.3cm) split
    NaI(Tl) annulus is described and experimental results are pre-
    sented as examples of its use in a number of coincidences and
    anti-coincidence experiments. The annulus as a whole was used
    as an Anti-Compton spectrometer and was found to reduce the
    Compton backgrounds in the Ge(Li) detector considerably. When
    the optically-isolated halves of the annulus were used separate-
    ly, the system was found to be very efficient for triple-coin-
    cidence experiments and as a pair spectrometer.

51. Auchampaugh, George F. SEARCH FOR DIRECT NEUTRON CAPTURE IN
    Cu63, Co59 AND Mn55 USING AGE(LI) DETECTOR. California
    University, Livermore, Lawrence Radiation Lab. Nov. 15,
    1968. 61p.   UCRL-50504

    An 18cc Ge(Li) detector was used to measure the intensity of
    5 transitions in copper, 9 in cobalt, and 11 in manganese as a

function of neutron energy.

52. Avida, R.; Atzmony, U., and Unna, I. FINITE SOLID ANGLE CORREC-
    TIONS TO ANGULAR CORRELATIONS FOR SQUARE GE(LI) GAMMA DETEC-
    TORS. <u>Nucl. Instrum. Methods</u>, 46 (1967), 350-354.

    Geometrical corrections for angular correlation measurements
    with square Ge(Li) detectors and extended linear sources are
    calculated.

53. Avignone, III, F. T. and Frey, G. D. ANGULAR CORRELATION AND
    DISTRIBUTION ATTENUATION COEFFICIENTS FOR PLANAR AND CO-
    AXIAL GE(LI) DETECTORS. <u>Rev. Sci. Instrum.</u>, 39 (Dec. 1968),
    1941-3.

    The coefficients Q_2 and Q_4 used to correct experimentally
    determined A_2 and A_4 angular correlation and distribution coef-
    ficients for the finite solid angle subtended by the detectors,
    are tabulated for various planar and coaxial Ge(Li) detectors.

54. Baertsch, R. D. and Hall, R. N. GAMMA RAY DETECTORS MADE FROM
    HIGH PURITY GERMANIUM. <u>IEEE Trans. Nucl. Sci.</u>, NS-17,
    no. 3 (June 1970), 235-40.
                                        Also report CONF-700301

    From the 12th Scintillation and Semiconductor Counter Sym-
    posium, Washington, D. C.

    A solution regrowth technique for growing P+ and N+ contacts
    on high purity germanium is described. Copper contamination of
    the high purity germanium is minimized by using KCN to remove
    copper from the surface of the germanium and by the gettering
    ability of molten indium in contact with the germanium. Diodes
    with leakage currents as low as 3×10^{-11} amp for 2000 volts
    applied to a fully depleted 4mm thick detector have been fabri-
    cated. Preliminary measurements show that the resolution ob-
    tained with these diodes is comparable to the best Li drifted
    germanium detectors at 60 keV and 122 keV. Diodes have been
    warmed to room temperature as many as five times with no degra-
    dation in resolution.

55. Baertsch, R. D. PROGRESS IN THE FABRICATION OF γ-RAY DETECTORS
    FROM HIGH PURITY GERMANIUM. <u>IEEE Trans. Nucl. Sci.</u>, NS-18,
    no. 1 (Feb. 1971), 166-69.

56. Bailey, Norman A. and Kramer, Gordon. THE USE OF SEMICONDUCTOR
    LITHIUM DRIFTED P-I-N JUNCTION DETECTORS FOR DOSIMETRY.
    pp. 151-61 of "Symposium on Radiation Effects on Metals and
    Neutron Dosimetry." Phila., American Society for Testing
    and Materials, 1963.

    Applications of the lithium drifted p-i-n junction semi-
conductor radiation detector in dosimetry are explored.

57. Baldinger, E. and Buschor, F. LIFETIME MEASUREMENTS IN LITHIUM
    DRIFTED GERMANIUM P-I-N DIODES. pp. 13-24 of "The Proceed-
    ings of the Meeting on Special Techniques and Materials
    for Semiconductor Detectors, Ispra, Italy, 1968." Capellani,
    F. and Restelli, G., ed. June 1969, 296p.
                                        CONF-681049;AUR-4269

    A study of lithium drifted germanium p-i-n diodes. Review-
ing the lifetime of carriers injected into the intrinsic zone.

58. Baldinger, E. and Haller, E. MEASUREMENT OF CONVERSION COEFFI-
    CIENTS WITH A GE(LI) DIODE. Helv. Phys. Acta., 42 (Dec.
    1969), 944-56. (In German)

    High resolution Ge(Li)-detectors were used to register
simultaneously gammas and conversion electrons of nuclear transitions.

59. Baldinger, E. and Haller, E. METHOD OF EXAMINATION OF THE SUR-
    FACE OF A GE(LI) DIODE. Helv. Phys. Acta., 43, no. 2
    (Jan. 15, 1971), 295-6.

60. Baldinger, E. and Matile, G. ON THE PREPARATION OF LITHIUM-
    DRIFTED GE COUNTER DIODES. Z. Angew. Math. Phys., 16 (1965),
    822-826. (In German)

    The lithium-compensated Ge counters are used for the detec-
tion of α rays because of their excellent resolution power.
The fabrication of these counters and their characteristics are
described.

61. Baldwin, T. O. and Thomas, J. E. USE OF ANOMALOUS X-RAY TRANS-
    MISSION FOR THE DETECTION OF DEFECTS PRODUCED IN SILICON
    AND GERMANIUM BY FAST NEUTRON IRRADIATION. J. Applied Phys.,
    39, no. 9 (Aug. 1968), 4391-98.

Absolute x-ray intensities anomalously transmitted through fast-neutron-irradiated silicon and germanium crystals have been measured and compared with those transmitted through nearly perfect (unirradiated) crystals. The results indicate that the Borrmann intensities in both silicon and germanium irradiated with 4×10^{19} fast neutrons/cm^2 ($E > 0.6$ MeV) were reduced by nearly a factor of two for $\mu_0 t_0 \approx 30$. In contrast, very small (<5%) decreases in the Bragg intensities diffracted from the surfaces were observed. The observed decreases in intensity have been attributed to the strains in the vicinity of "disordered regions", previously observed in electron-microscopy studies of irradiated germanium. Isochronal annealing studies have demonstrated that the crystals could be restored to near their original state of perfection by heating to $\approx 700^{\circ}$C for 30 min. Two well-defined annealing peaks were found in both kinds of crystals, silicon annealing at about 190° and 630°C and germanium at about 150° and 400°C. These studies suggest that fast-neutron doses at ambient temperature in the range of 10^{19}–10^{20}/cm^2 ($E > 0.6$ MeV) are necessary for observations of appreciable effects on the x-ray intensities diffracted from silicon and germanium.

62. Baldwin, T. O.   EFFECT OF FAST NEUTRON IRRADIATION AT AMBIENT TEMPERATURE ON THE LATTICE PARAMETER OF SILICON AND GERMANIUM CRYSTALS. Phys. Rev. Letters, 21, no. 13 (Sept. 23, 1968), 901-3.

63. Balland, Jean-Claude; Pagnotte, Y.; Samueli, J., and Sarazin, A. PROBLEMS RELATED TO CHARGE COLLECTION VARIATIONS IN GE(LI) DETECTORS WITH SPECIAL REFERENCE TO TIMING. IEEE Trans. Nucl. Sci., NS-17, no. 3 (June 1970), 405-24.
                                                    Also report CONF-700301

From 12th Scintillation and Semiconductor Counter Symposium, Washington, D. C.

With the maximum likelihood method we have determined, in the case of Ge(Li) detectors for a random shape (but single interaction) and noise, the expression of the likelihood functions. If an electronic system could give this function as response to Ge(Li) pulses it would be the best pulse processor, but it is obviously impossible to realize. However the method gives the bounds of accuracy of measurements theoretically possible. For planar pulses, this bound is found to be proportional to the inverse of the expectation of the energy of the delivered current. For coaxial pulses, one derives bounds in low and high electric field regions. Results are not so simple as in

the previous case and additive terms appear in the formulas with
the exception of the energy of the current. Rather surprising
results are obtained: timing nearly as good for planar pulses
could be approached.

64. Balland, Jean-Claude. RESOLUTION EN TEMPS DES DETECTEURS AU
        GERMANIUM COMPENSE AU LITHIUM. (TIME RESOLUTI..Ƶ OF LITHIUM-
        COMPENSATED GERMANIUM DETECTORS.) Lyon Univ. Institut de
        Physique Nucleaire. Sept. 1967, 111p. (In French) Thesis
                                            Also report LYCEN-6766

        An attempt is made to develop a simplified formulation of
the problem of time measurements using Ge(Li) detectors by taking into consideration the physical characteristics of these
detectors and the limitations due to the associated electronic
instrumentation. The properties of these detectors and the
physical processes for gamma detection are reviewed. The electronic problems connected with energy resolution are then analyzed. Time measurements are studied theoretically. The instrumentation used is described. The amplification and discriminator circuits are given. Experimental results obtained in time
resolution are shown graphically.

65. Balland, Jean-Claude; Pigneret, J., and Samueli, J. J. SOME
        THEORETICAL RESULTS ON TIMING CHARACTERISTICS OF GE(LI)
        DETECTORS. Nucl. Instrum. Methods, 52 (1967), 351-353.

        The pulse shape resulting from the interaction of a gamma
ray with a thin Ge(Li) planar detector is calculated. It is
found to depend on one random variable. By means of a suitable
transformation, the distribution of the time at which the signal crosses a threshold is obtained. Some curves are presented.
(auth.)

66. Balland, Jean-Claude. THEORETICAL AND EXPERIMENTAL STUDIES ON
        THE TIME RESOLUTION OF GE(LI) DETECTORS. Lyon Univ., France.
        Institut de Physique Nucleare. April 1970. 93p. (In French)
        Thesis   Also report LYCEN-7023

67. Balland, Jean-Claude; Pigneret, J.; Samueli, J. J. and Sarazin,
        A. TIME RESOLUTION OF PLANAR GE(LI) DETECTORS IN $\gamma - \gamma$
        COINCIDENCE EXPERIMENTS. IEEE Trans. Nucl. Sci., 15, no. 1
        (Feb. 1968), 411-422.

        A theoretical study on the subject of planar Ge(Li) in four

parts. Part 1 discusses the calculation of charge and current pulses issued from photoelectric interaction with the use of the Ge(Li) detector. Part 2 through 4 discusses time resolutions, risetime, electronic noise and multiple interactions in a Monte-Carlo program.

68. Barker, J. R. and Hearn, C. J. THE OBSERVED TRAPPING PARAMETERS OF PHOTO-EXCITED CARRIED IN GERMANIUM AND SILICON. Phys. Letters, 26A (Jan. 15, 1968), 148-9.

It is proposed that anomalies in the Hall number are responsible for discrepancies between experimental and theoretical trapping parameters in germanium and silicon.

69. Barker, P. H. and Connor, R. D. ^{56}CO AS A CALIBRATION SOURCE UP TO 3.5 MeV FOR GAMMA RAY DETECTORS. Nucl. Instrum. Methods, 57 (1967), 147-151.

Energies and relative intensities of the transitions in ^{56}Fe which follow the decay of ^{56}Co were determined with Ge(Li) detectors and are presented. These results are combined with other recent high resolution studies to give the relative intensities of the principal ^{56}Co gamma rays with an accuracy of about 3%. This provides a means of obtaining the relative efficiency of a gamma ray detector in the energy range 800 to 3500 keV.

70. Barrette, J. and Monaro, S. NOTE ON THE ENERGIES OF THE GAMMA TRANSITIONS IN ^{113}Sn DECAY. Nucl. Instrum. Methods, 63 (1968), 235-236.

The energies of gamma rays emitted in the ^{113}Sn decay have been measured with a high resolution 30 cm^3 Ge(Li) detector. By intercomparing close-lying full energy peaks arising from suitable calibration sources as ^{57}Co, ^{203}Hg, ^{192}Ir and ^{68}Ge energy values for the gamma rays present in the ^{113}Sn decay have been determined as follows: 255:11±0.08 and 391.75±0.05 keV. Furthermore, the energies of some of the γ-transitions present in the decay of ^{133}Ba have been remeasured and yielded values of: 276.33±0.05, 302.85±0.05, 356.03±0.05, and 383.90±0.05 keV.

71. Battelle-Northwest, Richland, Wash. Pacific Northwest Lab. PLUTONIUM UTILIZATION PROGRAM. Technical Activities Quarterly Report, March-May 1969. July 1969. 70p.
BNWL-1106

72. Baum, J. J. and Helenberg, H. W. PLANAR PAIR GEOMETRY FOR GE(LI)
     DETECTORS. IEEE Trans. Nucl. Sci., NS-17, no. 1 (Feb. 1970),
     33-8.   Also report CONF-691017-Pt. 1

     From 16th Nuclear Science Symposium, San Francisco, Calif.

     Some of the more desirable features of large volume coaxial
     pair configuration of planar lithium drifted germanium diodes.
     The present work examines some of the more important considera-
     tions involved and presents data obtained using a 27 cm³ planar
     pair with energy resolution of 3.0 keV FWHM at 1.33 MeV.

73. Beghian, L. E.; Hofmann, F., and Wilensky, S. USE OF A NANO-
     SECOND TIME GATED LI-GE DETECTOR FOR MEASURING DE-EXCITATION
     GAMMA RAYS FROM FAST NEUTRON SCATTERING. Nucl. Instrum.
     Methods, 41 (1966), 141-3.

     Discusses the use of a Ge(Li) detector to observe de-excita-
     tion of gamma rays caused by nonelastic scattering of fast neu-
     trons.

74. Benoit, R. and Mandl, V. ON RESOLUTION MEASUREMENTS WITH GE(LI)
     DETECTORS AND RLC FILTER USING POLE-ZERO CANCELLATION TECH-
     NIQUE. Nucl. Instrum. Methods, 60 (1968), 121-22.

     A comparison between the conventional RCCR filter network
     and the RLC with pole-zero cancellation is presented. The re-
     sults obtained show an insufficient filtering of low frequencies
     by the RLC filter.

75. Béraud, R.; Berkes, I.; Daniere, J.; Escudie, B.; Levy, M., and
     Rougny, R. DECOMPOSITION OF PULSE HEIGHT SPECTRA OF GE(LI)
     DETECTORS. Nucl. Instrum. Methods, 60 (1968), 219-220.

     An analog decomposition is used to resolve overlapping peaks
     in a Ge(Li) spectrum. The decomposition is verified by a num-
     erical x^2 test.

76. Berg, R. E. and Kashy, E. MEASUREMENTS OF HIGH ENERGY γ-RAYS
     WITH GE(LI) DETECTORS. Nucl. Instrum. Methods, 39 (Jan.
     1966), 169-172.

     A method of measuring the energy of high-energy gamma rays
     is described. Parallel capacitors are used as an effective

energy multiplier of a pulser signal to compare unknown high
energy gamma rays to known low-energy standards. This method
has also provided an effective tool for measuring the linearity
of the energy response of Ge(Li) detectors. This response is
found to be linear to better than ±0.03% in the range 662 to
2614 keV and better than ±0.1% up to 6 MeV. As an example of
this capacitor multiplying method the energy of the gamma ray
for the 3-excited level of $^{16}0$ is measured to be 6127.8±1 .2 keV.

77. Bertolini, G.; Cappellani, F.; Fumagalli, W.; Henuset, M., and
        Restelli, G. LITHIUM DRIFTED SEMICONDUCTOR DETECTORS IN
        NUCLEAR SPECTROSCOPY. European Atomic Energy Community,
        Brussels, (Belgium). Aug. 31, 1965. 10 p.
                                                    EUR-2580.E

    A short description of the techniques used in the construc-
tion and some examples of the utilization of lithium drifted
semiconductor detectors in electron and gamma spectroscopy are
presented.

78. Bertolini, G. and Coche, A. SEMICONDUCTOR DETECTORS. New York:
        J. Wiley, 1968. 526p.

    Presents a study of the characteristics and applications of
semiconductor radiation detectors. Fundamental properties of
semiconductors, characteristics and construction of detectors,
electronics for energy measurements and time resolution, nuclear
radiation spectroscopy and search for semiconductor materials
for gamma ray spectroscopy are discussed.

79. Bertolini, G. USE OF SEMICONDUCTOR DETECTORS IN ACTIVATION
        ANALYSIS. Kerntechnik, 11 (Jan. 1969), 31-5.

    A description of the physical phenomena governing the detec-
tion process in a Ge(Li) counter is given. The detection efficiency and the peak-to-background ratio are discussed for different detector volumes and for special configuration. In particular the optimization of the energy resolution of the detector
plus associated electronics is analyzed. Finally the application to activation analysis is discussed for what concerns analytical sensitivity and high purity analysis. Future trends in
this field are presented.

80. Bertrand, F. E.; Peelle, R. W.; Love, T. A.; Fox, R. J.; Hill, N.,
        and Todd, H. A TOTAL-ABSORPTION DETECTOR FOR 60-MEV PROTONS

USING LITHIUM-DRIFTED GERMANIUM. <u>IEEE Trans. Nucl. Sci.</u>,
NS-13, no. 3 (June 1966), 279-284.

A lithium-drifted germanium diode has been used for total-
absorption detection for 59 MeV protons from Oak Ridge Isochro-
nous Cyclotron. The detector is 1.9 cm in diameter, has a depletion depth of 6± mm, is cooled to less than $85^{\circ}K$, and is
sealed in an aluminum can with a 0.0026-in.-thick window. The
diode was oriented so that the protons entered in a direction
parallel to the detector junction.

81. Bilaniuk, O.M. and Shafroth, S. M. PRODUCTION OF RADIOACTIVE
    SOURCES OF ALKALIMETAL AND ALKALINE-EARTH ISOTOPES. <u>Nucl.</u>
    <u>Instrum. Methods</u>, 57 (1967), 177-178.

    A method for producing sources of alkali-metal and alkaline-
    earth isotopes by Alpha bombardment of rare gases in a cell lined
    with removable mylar film is described. Gamma rays arising from
    the decay of ^{89}Rb and Sr sources absorbed on the mylar film were
    observed with a $17cm^3$ Ge(Li) detector.

82. Bilger, H. R. EXPERIMENTS ON GENERATION-RECOMBINATION NOISE
    IN GERMANIUM SINGLE CRYSTALS IRRADIATED WITH FAST NEUTRONS
    AND GAMMAS. pp. 160-6 of "Semiconductor Nuclear-Particle
    Detectors and Circuits." Brown, W. L., ed. Washington, D.C.,
    National Academy of Sciences, 1969.
                                    Also report CONF-670520

    From Conference on Semiconductor Nuclear Particle Detectors
    and Circuits, Gatlinburg, Tenn.

    Generation-recombination noise in germanium crystals was
    measured for a ^{60}Co γ source and for 14.7 MeV neutrons. It was
    found radiation damage changed the parameters but not the shape
    of the spectra. The changes could be removed by annealing.

83. Bilger, H. R. THE FANO FACTOR IN GERMANIUM AT $77^{\circ}K$. pp.50-62
    of "Semiconductor Nuclear Particle Detectors and Circuits."
    Brown, W. L., ed. Washington, D. C. National Academy of
    Sciences, 1969.   Also report CONF-670520

    From Conference on Semiconductor Nuclear Particle Detectors
    and Circuits, Gatlinburg, Tenn.

    Pulse-height spectra for several gamma sources were measured

with a Ge detector at 77°K and the Fano factor was calculated
from the data. A value of 0.129 was obtained and it was con-
cluded that the Fano factor is essentially energy-independent
over the range considered.

84. Bilger, H. R. and Sherman, I. S. HIGH-RESOLUTION PHOTON SPEC-
    TROMETRY WITH LITHIUM DRIFTED GERMANIUM DETECTORS. Phys.
    Letters, 20 (Mar. 15, 1966), 513-16.

    Line widths of 850 ev(f.w.h.m.) were obtained for γ-rays
    below 100 keV with new field-effect transitions in the pre-
    amplifier used with germanium detectors. The "doublet" of
    ^{133}Ba (cascade γ rays from ^{133}Cs) at 80 keV, which consists of
    two lines 1.5 keV apart, was partially resolved.

85. Bilinski, J. R.; Brooks, E. H.; Cocca, U.; and Maier, R. J.
    PROTON NEUTRON DAMAGE EQUIVALENCE IN SI AND GE SEMICONDUCTORS.
    IEEE Trans. Nucl. Sci., NS-10 (1963), p. 71.

86. Bird, J. R.; Kenny, M. J., and Allen, B. J. NEUTRON RESONANCE
    MEASUREMENTS WITH A GERMANIUM DETECTOR. Phys. Letters,
    27B (Oct. 14, 1968), 638-40.

    A Ge(Li) detector is discussed with respect to its suitabili-
    ty for observation of resonances and measurement of total capture
    of neutron with energies greater than 5 keV.

87. Black, J. L. and Gruhle, W. CALCULATION OF ANGULAR CORRELATION
    ATTENUATION FACTORS AND EFFICIENCIES FOR LITHIUM-DRIFTED
    GERMANIUM DETECTORS. Nucl. Instrum. Methods, 46 (1967),
    213-222.

    Angular correlation attenuation factors and efficiencies
    have been computed for eleven Ge(Li) detectors with sensitive
    volumes in the range $0.3cm^3$ to $8.0cm^3$. The calculations were
    carried out for eight source-detector separations in the range
    1.0cm to 10.0cm.

88. Black, W. W. APPLICATION OF CORRELATION TECHNIQUES TO ISOLATE
    STRUCTURE IN EXPERIMENTAL DATA. Nucl. Instrum. Methods,
    71 (1969), 317-27.

    A method is presented for automatically locating the struc-
    ture, e.g. peaks, in experimental data through application of
    the cross-correlation function. Advantages of the method are

simplicity, insensitivity of parameters, and a small number of
parameters. To illustrate the versatility of this approach, example analyses are presented of data obtained with lithium-
drifted germanium detectors, lithium-drifted silicon detectors,
sodium-iodide detectors, and from neutron time-of-flight analyzers. Some comments are given on the general application of
the method as it relates to automatic data analysis, particularly with on-line computers.

89. Blankenship, J. L. DESIGN OF LOW NOISE VACUUM-TUBE PULSE AMPLI-
    FIERS FOR SEMICONDUCTOR RADIATION-DETECTOR SPECTROSCOPY.
    IEEE Trans. Nucl. Sci., NS-11, no. 3 (June 1964), 373-381.

     The pulse height resolution of semiconductor nuclear radia-
tion detectors for particle energies below 200 keV is primarily
limited by detector and amplifier noise. Amplifier noise may
be large compared to that contributed by cooled, deep-depletion
layer diodes, and minimum noise is achieved by optimizing amplifier design for specific diode parameters.

90. Blankenship, J. L. and Borkowski, C. J. IMPROVED TECHNIQUES
    FOR MAKING P-I-N DIODE DETECTORS. IRE Trans. Nucl. Sci.,
    NS-9, no. 3 (June 1962), 181-9.

     Techniques for achieving a thin dead layer exhibiting low
sheet resistance on the n^+ side of a p^+-i-n - diode made by the
lithium drift process were developed. A controlled quantity of
lithium was diffused through a 1- to 2-micron phosphorus-doped
layer on the silicon diode. The phosphorus-doped layer provided low sheet resistance. Because most of the lithium-diffused
layer was drifted into the bulk material, dead layers of less
than 7 microns thickness were achieved. Detectors made by this
technique have given 23 keV (fwhm) resolution for gamma rays and
monoenergetic electrons at room temperature, limited by diode
noise. Detectors cooled to 78 to 195°K gave 6.5 keV resolution
for Cs^{137} conversion electrons (625, 655 keV) and Pb^{207} x rays
(74, 90 keV). Detectors stored without bias voltage at room
temperature did not change performance over a four-month period.
An analysis of the drift parameters show that the lithium drift
rate depended upon the power dissipated in the diode during
drift. An automatic control system was developed which allows
the lithium drift operation to proceed at power dissipations in
excess of 50 watts.

91. Blankenship, J. L. and Borkowski, C. J. USE OF LITHIUM-DRIFTED
    DIODES IN BETA AND GAMMA SPECTROMETRY. Rev. Sci. Instrum.,
    33 (July 1962), 778-80.

The performance of Li–drifted p^+–i–n^+ diode detectors, which have thin dead layers and deep sensitive volumes, is described for beta, x-ray and gamma spectrometry at energies below 100 keV. The use of these detectors in measurements of the electron and photon spectra of Bi^{207}, Pb^{207}, and Cs^{137} is shown.

92. Bolotin, H. H. GAMMA-RAY COINCIDENCE TECHNIQUES EMPLOYING HIGH-RESOLUTION GERMANIUM DETECTORS. Argonne National Lab. III. Phys. Div. Report. April 1966.

The technique of coincidence measurements with high resolution lithium drifted germanium diode detectors is described. It is being used to investigate the highly complex gamma ray spectra following capture of slow neutrons. Information obtained on the properties of low-lying excited states of off-odd nuclei is illustrated by recent results on Ga^{72} from neutron capture in Ga^{71} and on Ga^{66} from beta decay of Ga^{66}. The lithium-drifted germanium diode detectors combine extremely high energy resolution, moderate detection efficiency, and reasonably fast coincidence timing capabilities. The typical complexity of gamma ray spectra following slow neutron capture is shown by the spectrum of low energy transitions from the reaction Ga^{71} $(n\gamma)Ga^{72}$, as viewed by a lithium-drifted germanium diode detector.

93. Bolotin, H. H. USE OF GERMANIUM(LITHIUM) DETECTORS IN GAMMA-RAY SPECTROSCOPY: APPLICATIONS AND BASIC CONSIDERATIONS. pp. 660-83 of "Semiconductor Nuclear-Particle Detectors and Circuits." Brown, W. L., ed. Washington, D. C., National Academy of Sciences, 1969.   Also report CONF-670520

From Conference on Semiconductor Nuclear Particle Detectors and Circuits, Gatlinburg, Tenn.

A review of the fundamental advantages, and limitations on the use of Ge(Li) detectors for gamma singles and coincidence spectroscopy. Experimental applications are described. Also some extremely valuable experiments are described which are difficult or unfeasible because of detector limitations.

94. Boonstra, Alexander Hendrik. SOME INVESTIGATION ON GERMANIUM AND SILICON SURFACES. Technische Hogeschool, Eindhoven, (Netherlands). June 1967. 118p. Thesis

The results of theoretical and experimental investigations on the surface properties of germanium and silicon single crystals are discussed.

95. Bornand, Bernard. METHODS OF FABRICATION FOR GERMANIUM AND SIL-
    ICON DETECTORS WITH COMPENSATION BY LITHIUM-ION DRIFTING.
    (METHODS DE FABRICATION DE DETECTEURS AU GERMANIUM ET AU
    SILICIUM COMPENSES PAR MIGRATION D'IONS DE LITHIUM). March
    1968. 74p. (In French)   ICEA-BIB-90

    A bibliography on the fabrication of germanium and silicon
detectors with compensation by lithium-ion drifting is presented.
Mounting, packaging, encapsulation, and methods of control for
preparation are included, as also resolution and efficiency
measurements and spectrometry when the fabrication is concerned.
A list of abstracts with references to reviews, reports, thesis,
proceedings, and patents (168 references) from 1964 to 1967
(1st semester) with some references at earlier dates (e.g. patents) is given. There are included subject index, periodical,
report and patent number index and personal author index. In a
supplement, 12 abstracts of communications for the 11th Scintillation and Semiconductor Counter Symposium (Washington, Feb. 28-
March 1, 1968) are presented.

96. Borkowski, C. J. and Fox, R. J. GAMMA-RAY COMPENSATION OF
    GERMANIUM. pp. 333-336 of "Semiconductor Nuclear-Particle
    Detectors and Circuits." Brown, W. L., ed. Washington,
    D. C., National Academy of Sciences, 1969.

    Discusses the performance of a germanium detector compensa-
ted with acceptor defects produced by ^{60}Co gamma rays. The advantages of gamma compensation over lithium compensation for
germanium are discussed.

97. Bosch, H. E. and Szichman, E. DETERMINATION OF THE THREE CON-
    VERSION COEFFICIENTS IN THE DECAY OF AU^{198} WITH SEMICONDUC-
    TOR COUNTERS. Instituto de Investigacion Aeronautice y Es-
    pacial, Buenos Aires, Argentina. Dec. 1966. 18p.
                                            LR-15; AD-648754

    A set-up for determining conversion coefficients by means of
different arrangements of semiconductor detectors (both Lithium-
drifted silicon and germanium) and scintillation counters is
presented. The conversion coefficients corresponding to the
three transitions in ^{198}Hg were measured. Single electron conversion and gamma spectra were recorded, deriving the value of
the conversion coefficient for the 1.083 MeV transition, taking
into account the value obtained for that of the 0.412 MeV transition. Beta continuum-electron conversion coincidences have
been performed with two lithium-drifted silicon detectors and
the K-conversion coefficient for the latter transition was obtained. Gamma-gamma coincidences with a lithium-drifted germanium

detector and a conventional scintillation counter were done,
yielding the value of the K-conversion coefficient for the
0.671 MeV transition.

98. Bosch, H. E.; Haverfield, A. J.; Szichman, E., and Abecasis, S. M.
    HIGH-RESOLUTION STUDIES IN THE DECAY OF ^{133}Ba WITH SEMI-
    CONDUCTOR COUNTERS. Nucl. Phys., A108 (1968), 209-20.

    A reinvestigation of some properties of the decay of ^{133}Ba
    was carried out with high-resolution equipment based on Li-drifted
    silicon and germanium semiconductor counters. Electromagnetic
    transitions of 53.4, 79.7, 80.9, 160.6, 223.1, 276.5, 302.8,
    356.2 and 383.7 keV were observed and a level sequence of ener-
    gies -0.9, 160.6, 383.7 and 437.1 keV was derived. The K-con-
    version coefficients for the above transitions were determined
    either by electron-gamma and gamma-gamma coincidences or by
    comparison of the intensities of the electron-conversion and
    gamma-ray peaks of the corresponding "singles" spectra; K/LM
    ratios are also determined. The electron-capture branching
    ratios to the four excited levels of ^{133}Cs as well as lot ft
    values were derived. A decay scheme is proposed.

99. Brandenberger, J. D. TIME MEASUREMENTS WITH GE(LI) DETECTORS.
    Nucl. Instrum. Methods, 69 (Apr. 1969), 271-6.

    A method of obtaining improved time resolution from coin-
    cidence measurements utilizing large volume Ge(Li) detectors
    is presented. The analysis indicates that a time resolution of
    the order of 1 ns is obtainable if a preamplifier with a short
    rise time is used. An experimental time resolution of 3.8 ns
    is obtained using a preamplifier with a rise time considerably
    longer than the charge collection time constant of a Ge(Li) de-
    tector. These results all apply to detection of gamma rays hav-
    ing a large range of energies detected with a large volume co-
    axial detector.

100. Brashear, H. R. and Shipp, R. L. CRYOSTATS FOR LITHIUM-DRIFTED
     DIODE DETECTORS. p. 31 of "Oak Ridge National Lab. Instrum-
     entation and Controls Division. Annual Progress Report,
     Sept. 1, 1967." 1967.   ORNL-4219

101. Braun, H. PRODUCTION OF COUNTER DIODES ACCORDING TO THE LITHIUM-
     DRIFT METHODS. Acta. Phys. Aust., 23 (1966), 393-6. (In
     German)

     A method for the preparation of β counters from Si using the

Li drift method is described. Because of the low cost the
apparatus is especially suitable for laboratory applications.
The principle of the Li drift method is given.

102. Brethon, Jean-Pierre; Libs, Gerard; DeTourne, Guy; Legrand, Jean,
       and Boulanger, Jean-Pierre. SPECTROMETRIE γ A HAUTE PER-
       FORMANCE A L'AIDE DE DETECTEURS GE-LI. (HIGH PERFORMANCE
       γ SPECTROMETRY USING GE(LI) DETECTORS.) Commissariat a
       l'Energie Atomique, Saclay, France. Centre d'Etudes Nucleares.
       Jan. 1968. 12p. (In French)   CEAR-R-3405

        A high resolution gamma spectrometer design which uses Ge(Li)
   detectors, a cooled field effect transistor preamplifier, and
   a spectrum stabilizer is described. The obtained resolution
   for the ^{122}keV gamma ray of ^{57}Co is 0.96 keV, and ^{239}Pu, ^{233}Pa,
   and ^{95}Zr and ^{95}Nb spectra are shown for the example.

103. Broder, D. L.; Gamalii, A. F.; Lashuk, A. I.; Nesterov, B. V.,
       and Sadokhin, I. P. CROSS SECTIONS OF THE FORMATION OF GAMMA
       QUANTA IN THE REACTION (n, n' gamma) IN FLOURINE, COBALT,
       ANTIMONY AND TANTALUM NUCLEI. (SECHENIYA OBRAZOVANIYA
       GAMMA-KVANTOV V REAKTSII (n, n GAMMA) NA YADRAKH FTORA,
       KOBALTA, SUR'MY I TANTALA). Gosudarstvennyi Komitet po
       Ispolzovaniyu Atomnoi Energii, Obninsk (USSR) Fiziko-Ener-
       geticheskii Inst. 1969. 24p. (In Russian)   FEI-155

104. Broman, L. and Dubois, J. PRECISION DETERMINATION OF ENERGY
       LEVELS IN ^{43}Sc USING A LITHIUM-DRIFTED GERMANIUM SOLID STATE
       DETECTOR. Ark. Fys., 30 (1965), 498-99.

105. Broman, L. A SPECTROMETER FOR NEUTRON CAPTURE GAMMA RAY STUDIES.
       Nucl. Instrum. Methods, 50 (1967), 29-37.

        Due to the large Compton background from intense gamma lines
   the usefulness of the Ge(Li) detector for the investigation of
   neutron capture gamma ray spectra between 500 and 2000 keV is
   restricted to the determination of strong gamma lines. By placing the Ge(Li) detector in the reflected beam from a single crystal spectrometer this background can be completely suppressed
   within a variable energy interval. This combination has been
   tested on the complex gamma yield from thermal neutron capture
   in ^{167}Er. When using the Ge(Li) detector as a photo effect
   spectrometer gamma rays up to about 1500 keV have been detected.
   The energy resolution of the combined spectrometer, using this
   particular Ge(Li) detector is better than that of double flat
   and curved crystal spectrometers above about 700 keV. Recent

developments in solid state detector techniques will appreciably lower this limit.

106. Bromley, D. A.   NUCLEAR EXPERIMENTATION WITH SEMICONDUCTOR DETECTORS.   IRE Trans. Nucl. Sci., NS-9, no. 3 (June 1962), 135-54.

Significant progress has been made during the past year toward realizing the very great potential of semiconductor detectors in nuclear research; this progress is reviewed, together with brief reference to relevant earlier work. Representative fields where major advances have been recorded are those of electron and high energy heavier charged particle detection using lithium-drifted and other thick depletion layer devices, neutron detection studies, nuclear astrophysical research where the low mass, size, and power requirements are of paramount importance, focal plane studies with magnetic spectrographs involving multi-channel semiconductor systems, nuclear reaction product isotope identification with both hybrid gas and semiconductor detectors and with all semiconductor systems, high efficiency particle detection with large area mosaic detectors, and fission and other studies in high ambient neutral radiation fields where the junctions have unique advantages.

107. Brooks, W.; Magdics, A.; Molen, G., and Rosh, D.   PLUTONIUM PASSIVE ASSAY FACILITY PLANT INSTRUMENTATION PROGRAM.   First Quarterly Progress Report, April-June 1970.   United Nuclear Corp., Elmsford, N.Y. Aug. 14, 1970. 25p.   BHO-66-1

108. Brooks, W. and Rosh, D.   PLUTONIUM PASSIVE ASSAY FACILITY PLANT INSTRUMENTATION PROGRAM.   Second Quarterly Progress Report, July-September 1970.   United Nuclear Corp., Elmsford, N.Y. Research and Engineering Center. Nov. 30, 1970. 5p.
                                                                      BHO-66-2

109. Broude, C.   GAMMA-RAY ANGULAR CORRELATIONS FOLLOWING NUCLEAR REACTIONS.   Chalk River Nuclear Labs., Ont. 1965. 15p.
                                                           CONF-651118-6

From American Physical Society Meeting, Gatlinburg, Tenn.

New developments in the techniques of measuring gamma angular correlations following capture reactions are considered. The methods discussed for preparing discrete gamma-emitting states in an axially symmetric way for analysis of measured gamma correlations are via an intermediate unobserved radiation.

110. Broude, C.; Haeusser, O.; Malm, H.; Sharpey-Schafer, J. F., and
     Alexander, T. K. A GE(LI) TWO CRYSTAL COMPTON SPECTROMETER.
     Nucl. Instrum. Methods, 69 (Mar. 1969), 29-34.

     A performance of Compton spectrometers using two Ge(Li)
     gamma-ray detectors is demonstrated. Two different configura-
     tions of detectors have been used together with on-line computer
     methods to show that a nearly-single-peak spectrum of good re-
     solution can be obtained for a monochromatic gamma ray. Further,
     the peak is due to purely single Compton scattering of well de-
     termined scattering angle despite the large solid angles of the
     scattering and detecting crystals. These considerations make
     the device well suited to use as a gamma ray polarimeter.

111. Broude, C.; Sharpey-Schafer, J. F.; Haeusser, O., and Alexander,
     T. K. A GERMANIUM (LITHIUM) TWO CRYSTAL COMPTON SPECTROME-
     TER. Atomic Energy of Canada, Ltd., Chalk River, Ontario.
     pp. 688-92 of "Semiconductor Nuclear-Particle Detectors and
     Circuits." Brown, W. L., ed. Washington, D. C., National
     Academy of Sciences, 1969.   Also report Conf-670520

     From Conference on Semiconductor Nuclear Particle Detectors
     and Circuits, Gatlinburg, Tenn.

     A Compton spectrometer is described which uses two Ge(Li)
     detectors and which may be used as a polarimeter. The gamma
     spectrum of ^{88}Y was measured to check the performance of the
     spectrometer.

112. Brownridge, James and McLoughlin, David. TECHNIQUE FOR DETER-
     MINATION OF DEPLETION DEPTH IN GE(LI) COUNTERS. Nucl.
     Instrum. Methods, 60 (1968), 116-120.

     A method for determining the extent of the compensated re-
     gion of P-type germanium drifted with lithium is described.
     It is a non-destructive method which is applied while the cry-
     stal is in the drifting apparatus and does not interrupt the
     drifting process.

113. Brune, D.; Dubois, J., and Hellstroem, S. IMPROVEMENTS IN
     APPLIED GAMMA-RAY SPECTROSCOPY BY GERMANIUM SEMI-CONDUCTOR
     DETECTOR. Nukleonik, 7 (Oct. 1965), 484-8.

     A germanium semiconductor detector was used in the investi-
     gation in four cases of applied gamma-ray spectroscopy.

114. Buck, T. M. SURFACE NOISE IN SEMICONDUCTOR DETECTORS. pp. 144-55
     of "Semiconductor Nuclear-Particle Detectors and Circuits."
     Brown, W. L., ed. Washington, D. C., National Academy of
     Sciences, 1969.   Also report CONF-670520

          From Conference on Semiconductor Nuclear Particle Detectors
     and Circuits, Gatlinburg, Tenn.
          Evidence for a simple model which has been useful in under-
     standing and controlling surface-generated noise in semiconduc-
     tor detectors is reviewed. Additional noise mechanisms that
     have come to light are discussed.

115. Bueker, H. HIGH RESOLUTION GAMMA SPECTROMETER WITH A BOREHOLE
     GERMANIUM DETECTOR. Nucl. Instrum. Methods, 69 (Apr. 1969),
     293-302. (In German)

          Description is given for a high-volume lithium-drifted bore-
     hole germanium detector and the vacuum chamber in which it is
     operated. The detector is especially suited for investigations
     of low gamma-active samples often arising in the fields of med-
     icine, biology and radiochemistry. The germanium crystal is
     built into a vacuum-tight casing on which an ion-spray-pump is
     welded, and by which the detector casing can further be evacu-
     ated in case of necessity.

116. Bueker, H. HIGH RESOLUTION GAMMA-SPECTROMETRY WITH LITHIUM-
     DRIFTED GERMANIUM DETECTORS. PART 1. Kerntechnik, 10
     (Dec. 1968), 687-94. (In English and German)

          The structure and operation of a lithium-drifted germanium
     detector are discussed. The generation of an output pulse is
     described. A comparison is also made between Ge(Li) and NaI
     detectors for gamma spectroscopy. The design of typical detec-
     tor chamber for large volume Ge detectors is examined.

117. Bueker, H. HIGH-RESOLUTION GAMMA SPECTROMETRY WITH LITHIUM-
     DRIFTED GERMANIUM DETECTORS. PART 2. Kerntechnik, 11
     (Jan. 1969), 39-45. (In English and German)

          A description of the basic electronic components of lithium-
     drifted germanium gamma spectrometers is presented including
     the pre-amplifier, the linear amplifier, and the multi-channel
     analyzer. Also, the use of germanium gamma spectrometers in
     neutron activation analysis and fission products investigations
     is outlined.

118. Buhler, S. and Marcus, L.   COLD FINGER CRYOSTATS FOR LITHIUM-
DRIFTED GERMANIUM DETECTORS. <u>Nucl. Instrum. Methods</u>, 50
(1967), 170-72.

Cold finger type cryostats of different shapes with self-
containing vacuum are described. Very long cold fingers need
internal liquid nitrogen circulation in order to shorten the
cool-down period and to get lower temperatures of the detector.
Conditioning and performances of such cryostats are described.

119. Bullis, W. M. and Coleman, J. A.   CHARACTERIZATION OF GERMANIUM
AND SILICON FOR NUCLEAR RADIATION DETECTORS. pp. 166-75 of
"Nucleonics in Aerospace." Polishuk, Paul, ed. New York,
Plenum Press, 1968.   Also report CONF-670714

From 2nd International Symposium on Nucleonics in Aerospace,
Columbus, Ohio.

Uncontrolled variations in the quality of germanium and silicon for radiation detection have occurred for reasons which are
not well understood. Crystals are presently selected on the
basis of room temperature resistivity, photoconductive decay
lifetime and etch pit density. Use of these parameters does not
always enable one to discriminate between material suitable for
detector fabrication and material which is not.

120. Bullis, W. and Baroody, A. J., Jr., eds. METHODS OF MEASUREMENT
FOR SEMICONDUCTOR MATERIALS, PROCESS CONTROL, AND DEVICES.
Quarterly report, April 1–June 30, 1970. National Bureau
of Standards, Washington, D. C. Nov. 1970. 58p.
NBS-TN-560

NBS activities directed toward the development of methods
of measurement for semiconductor materials, process control, and
devices are described. Principal emphasis is placed on measurement of resistivity, carrier lifetime, and electrical inhomogeneities in semiconductor crystals; evaluation of wire bonds,
metallization adhesion, and die attachment; and measurement of
thermal properties of semiconductor devices and electrical properties of micro-wave devices. Work on related projects on
silicon nuclear radiation detectors and specification of germanium for gamma-ray detectors is also described. Supplementary
data concerning staff, standards committee activities, technical
services, and publications are included as appendixes.

121. Buschmann, H. T. and Lauterjung, K. H.   THE GAMMA SPECTRUM OF
^{214}PO. <u>Z. Phys.</u>, 207 (1967), 411-27. (In German)

A high-resolution gamma-ray spectrometer with a Ge(Li) detector is described. The spectrometer is used as a single-crystal
spectrometer as well as a three-crystal pair spectrometer with
a Ge(Li) central detector and two Na-crystals for the detection
of the annihilation quanta. Measurements were made of ^{214}Po.
The sources were ^{222}Rn small glass tubes. 66 gamma rays, 10 of
which were previously unknown, were observed. An improved level
scheme of ^{214}Po is given.

122. Cacheux, J.; Fertin, J., and Meuleman, J. CONSTRUCTION AND
     CHARACTERISTICS OF LITHIUM-COMPENSATED GERMANIUM DETECTORS.
     Ind. At., 12, no. 3-4 (1968), 83-93. (In French)

     After a very brief summary of the interactions of electro-
     magnetic rays with forms of matter, a listing is provided of
     the advantages obtainable by the use of monocrystalline ger-
     manium compensated by lithium. The properties of the p-i-n
     junctions resulting are then studied. The various techniques
     for the presentation of these junctions when cooled to a very
     low temperature are next examined, as well as the electronics
     involved, and the utilization of the signals furnished. Lastly,
     the prospects of a cell of universal conditioning is considered.

123. California University, Lawrence Radiation Lab. HAZARDS CONTROL
     QUARTERLY REPORT, NO. 24. Jan.-March 1966. 14p.
                                                          UCRL-14881

     Short summaries of the research work at Lawrence Radiation
     Laboratory are represented. This report describes briefly the
     research in the area with lithium-drifted Ge gamma detectors.

124. California University, Lawrence Radiation Lab. LITHIUM DRIFTED
     GERMANIUM DETECTORS (ENGINEERING MATERIALS). 18 DRAWINGS.
                                                          CAPE-1317

     Eighteen drawings of lithium drifted germanium detectors and
     their use in high resolution and spectroscopy in nuclear decay
     scheme determinations, and in the identification of isotopic
     contents are presented. Two types of detectors are discussed.

125. Camp, David C. APPLICATIONS AND OPTIMIZATION OF THE LITHIUM
     GERMANIUM DETECTORS. California University, Lawrence Radi-
     ation Lab. March 1967. 102p.   UCRL-50156

     The concept of semiconductors and how the semiconductor ma-
     terials silicon and germanium are used as radiation detectors

are discussed. The various types of lithium—drifted germanium
detectors that have evolved are also discussed. Some of the
requirements necessary if the detector and the electronic components that comprise the total system are to give the best
possible performance are presented.

126. Camp, David C. COMPTON SUPPRESSION AND PAIR SPECTROMETER.
     pp. 693-702 of "Semiconductor Nuclear-Particle Detectors
     and Circuits." Brown, W. L., ed. Washington, D. C.,
     National Academy of Sciences, 1969.
                                        Also report CONF-670520

     From Conference on Semiconductor Nuclear Particle Detectors
and Circuits, Gatlinburg, Tenn.

     A gamma spectrometer is described which utilizes a Ge(Li)
detector partially enclosed by NaI(Tl) scintillation detectors
to suppress the Compton distribution. The arrangement may be
also used as a pair spectrometer.

127. Camp, D. C. and Van Lehn, A. L. CORRECTIONS TO COAXIAL GE(LI)
     DETECTOR SOLID ANGLE CORRECTION FACTORS*. Nucl. Instrum.
     Methods, 87 (1970), 147-48.

     This letter discusses additional corrections needed for
solid angle correction factors for coaxial detectors.

128. Camp, D. C. and Van Lehn, A. L. FINITE SOLID-ANGLE CORRECTIONS
     FOR GE(LI) DETECTORS. Nucl. Instrum. Methods, 76 (1969),
     192-240.

     Finite solid-angle corrections have been calculated for Ge(Li)
detectors. Planar detectors considered vary from 4 to 20 cm^2
and from 6 to 16 mm in depletion depth. True circular coaxial
detectors vary from 26 to 50 mm in diameter with depletion depths
from 7 to 20 mm and lengths from 22 to 57 mm. Five-sided detectors of both trapezoidal and circular cross sections are included, and these range from 20 to 70 mm in length and from 8
to 20 mm in depletion depths. The corrections are calculated
for source distances of 3.0, 5.0, 7.0, and 10.0 cm.

129. Camp, D. C. and Armantrout, G. A. LITHIUM-DRIFTED GERMANIUM
     DETECTORS FOR HIGH-RESOLUTION BETA- AND GAMMA-RAY SPECTRO-
     SCOPY. Lawrence Radiation Lab., University of California.
     March 30, 1965. 31 p.
                         UCRL-12245; also report CONF-650507-2

From IAEA Symposium on Radioisotope Sample Measurement Techniques in Medicine and Biology, Vienna.

Two types of germanium detectors have been fabricated using the lithium drift technique. Those of the first type have been active volumes in excess of 6 cm^3 and are primarily intended for high-energy (>1.0 MeV) γ spectroscopy. Those of the second type are large area, low capacity, windowless detectors intended for very-high-resolution B and low-energy spectroscopy. Both types are operated in a vacuum at liquid nitrogen temperature (77°K). The large volume detectors have areas greater than 6 cm^2 with depletion depths in excess of 1 cm. The experimental resolution (FWHM) obtainable with these detectors is limited at low energies by the noise level of the preamplifier, while at high energies (>1 MeV) the limitation is due to amplifier instability.

130. Camp, David C. NUCLEAR SPECTROSCOPY VIA GE(LI) DETECTORS IN COMPTON SUPPRESSION AND PAIR SPECTROMETERS. California University, Livermore, Lawrence Radiation Lab. Sept. 12, 1969. 92p.   UCRL-71825; Report CONF-690818-06

From International Conference on Radioactivity in Nuclear Spectroscopy, Nashville, Tenn.

In the six years that lithium-drifted germanium detectors have been available, the range of scientific disciplines and problems to which they have been applied has become vast and broad. The Ge(Li) detectors used for these experiments are primarily of four types and sizes, and these are briefly discussed. The relatively sonal volume of these detectors with their corresponding small peak-to-total and peak-to-Compton ratios has led to various compensations of Ge(Li), NaI(Tl), and plastic scintillation detectors which are designed to reduce the Compton distribution. The resulting configurations, their relative performance and cost is discussed.

131. Campbell, J. L.; Smith, H. J., and McKenzie, I. K. A COINCIDENCE TECHNIQUE FOR STUDY OF GE(LI) DETECTOR PROFILES. Nucl. Instrum. Methods, 92, no. 2 (March 15, 1971), 237-45.

132. Campbell, J. L.; O'Brien, P., and McNelles, L. A. EFFICIENCY CALIBRATION OF SEMICONDUCTOR X-RAY DETECTOR. Nucl. Instrum. Methods, 92, no. 2 (1971), 269-75.

Data on the decay characteristics of several radioactive sources suitable for efficiency calibrations of Ge(Li) and Si(Li) detectors in the energy range 1 to 100 keV are compiled and discussed.

133. Campbell, J. L.; Goble, R. J., and Smith, H. J. EFFICIENCY OF
      PLANAR GE(LI) DETECTORS AT VERY LOW ENERGY. Nucl. Instrum.
      Methods, 82 (1970), 183-6.

      From measurements of the intensities of collimated beams of
      6.4 and 14.4 keV photons in a thin-window Ge(Li) detector and a
      proportional counter, values of 0 to 35 A° and 0.3 um were ob-
      tained for the depths of the absorbing layers of gold and ger-
      manium in the central region of the former. It is shown that
      reliable determination of the Ge(Li) detector efficiency at low
      energy (88.3-0.3 O/O at 6.4 keV in this case) requires precise
      measurement of various absorption coefficients whose tabulated
      values differ considerably.

134. Cappellani, F.; Fumagalli, W.; Henuset, M., and Restelli, G.
      FABRICATION OF GERMANIUM LI DRIFTED DETECTORS. Nucl.
      Instrum. Methods. 47 (1967), 121-124.

      A description of the procedures used in constructing Ge(Li)
      detectors is given.

135. Cappellani, F.; Fumagalli, W., and Restelli, G. GERMANIUM COM-
      PENSATION BY LI-DRIFT IN BOILING LIQUIDS FOR P-I-N DIODE
      CONSTRUCTION. Nucl. Instrum. Methods, 37 (1965), 354-6.

      Fabrication of lithium-drifted germanium p-i-n diodes hav-
      ing compensated volumes as high as 4 cc is reported. The li-
      thium drifting process is carried out in boiling liquids in or-
      der to provide temperature control and heat removal. The liquids
      studied include pentane, hexane, o-xylene, chloroform, methenol,
      acetone, and petroleum spirits.

136. Cappellani, F.; Ostidich, A., and Restelli, G. INSENSITIVE ZONES
      IN THE INTRINSIC REGION OF GE(LI) COAXIAL DETECTORS. Nucl.
      Instrum. Methods, 79 (1970), 170-4.

      Insensitive regions in the intrinsic volume of a double open
      ended coaxial Ge(Li) detector have been put in evidence and eval-
      uated by using the gamma ray scanning technique with photons of
      different energy. Insensitive zones have been observed at both
      open ends of the diode, having circular geometry with a thick-
      ness increasing from the p core to the n+ electrode. Their size
      is sensitive to the detector bias and can be reduced by a suit-
      able warmup and by redrift cycles. Their existence is probably
      related to local imperfect compensation of the Li drifted ger-
      manium.

137. Capellani, F. and Restelli, G. PRELIMINARY RESULTS ON THE DETER-
MINATION OF IMPURITIES IN GERMANIUM. pp. 55-64 of "The Proceedings of the Meeting on Special Techniques and Materials
for Semiconductor Detector, Ispra, Italy, 1968." June 1969.
EUR-4269

138. Cappellani, F. and Restelli, G. PROCEEDINGS OF THE MEETING ON
SPECIAL TECHNIQUES AND MATERIALS FOR SEMICONDUCTOR DETECTORS,
ISPRA, ITALY, OCTOBER 1-2, 1968. European Atomic Energy
Community, Brussels, Belgium. Center for Information and
Documentation. June 1969. 296p.
EUR-4269; Also report CONF-681049

139. Carlson, George A. DIRECTIONAL CORRELATION ATTENUATION COEFFI-
CIENTS WITH GE DETECTORS. California Division, Crocker Nuclear Labs. Jan. 23, 1967. 32p. UCD-CNL-74

Methods are presented for finding accurate attenuation coefficients for a rectangular Ge detector. The coefficients and
the corresponding ones for a cylindrical NaI detector are used
in the Legendre polynomial series representing the directional
correlation function for the finite solid angles subtended by
the detectors. Examples are given for particular cases and the
computer program used in the calculation are included.

140. Casper, Karl J. FUNDAMENTAL STUDIES ON SEMICONDUCTOR RADIATION
DETECTORS. Western Reserve University, Cleveland, Ohio.
Annual Status Report, March 1, 1966-February 28, 1967. 1967.
93p. CSCL-09C; NASA-CR-84812

Details are presented on silicon and germanium lithium drifting apparatus, and the lithium drift process is described. A
computer program was written for gamma ray photopeak fitting to
aid in analysing gamma spectra measured with lithium-drifted germanium detectors. The program is included, and the processing
procedures are discussed. Further experiments using 22 and 160 MeV
protons for determining the response of large volume silicon
detectors are also reported.

141. Chalk River Nuclear Laboratory. See Atomic Energy of Canada, Ltd.

142. Chandra, Ramesh. ENERGY LEVELS IN [105]PD. Boston University,
Boston. 1968. 115p. Thesis

The decay of ^{105}Ag to the excited states of ^{105}Pd was investigated using high resolution solid state detectors. A singles spectrum with a Ge(Li) detector, coincidence spectra with
two Ge(Li) detectors, conversion coefficients for different
gamma rays and conversion electron spectra with a Si(Li) detector were measured. As a result, three new gamma rays of energy
54, 75, and 527 keV were identified. There is also a considerable improvement in the accuracy of the branching ratios and
the conversion coefficient.

143. Chapman, G. T. GAMMA-RAY ATTENUATION COEFFICIENTS FOR GERMANIUM.
    Nucl. Instrum. Methods, 52 (1967), 101-103.

An interpolation method based on the values listed in
NBS-583 for five elements adjacent to germanium was used to determine the gamma-ray attenuation coefficients of germanium.
If the region of the atomic numbers, Z, is relatively small and
the gamma-ray energy is not too close to the K-shell energies
for the elements in this range of Z, the attenuation coefficients will vary smoothly and not too drastically as a function
of Z.

144. Chartrand, M. G. and Malm, H. L. VARIATION OF THE GAMMA-RAY
    RESOLUTION OF GE(LI) DETECTORS WITH THE POSITION OF IRRA-
    DIATION? (VARIATION DE LA RESOLUTION PAR RAYONS GAMMA DES
    DETECTUERS AU GE(LI) SELON LA POSITION DE L'IRRADIATION?)
    Atomic Energy of Canada, Ltd., Chalk River (Ontario).
    June 1967. 28p.   AECL-2764

A study was made of the gamma-ray energy resolution of planar Ge(Li) p-i-n detectors showing varying degrees of charge
trapping. A small area collimated beam of ^{51}Cr (320 keV) gamma-
rays was used to irradiate the detectors both parallel and per-
pendicular to the junction surface. (For some measurements a
^{137}Cs source (662 keV) was also used.) Three detectors, from
different crystals, were studied in detail.

145. Chase, Robert L. PULSE TIMING SYSTEM FOR USE WITH GAMMA RAYS
    ON GE(LI) DETECTORS. Rev. Sci. Instrum., 39, no. 9 (Sept.
    1968), 1318-26.

A zero crossing technique has been employed to reduce energy-
dependent time errors, when detecting gamma rays with Ge(Li)
detectors. The timing errors associated with pulse shape variation have been evaluated so that circuit parameters might be
chosen to minimize those errors. Results of coincidence measurements using a plastic scintillator and a Ge(Li) detector are

presented. Typical results using an 8cc planar detector are
0.8 nsec fwhm at 2.75 MeV, 2.8 nsec fwhm at 511 keV, and 1.5 nsec
variation in coincidence peak position from 500 keV to 2.75 MeV.

146. Chasman, C. and Ristinen, R. A. A CRYOSTAT FOR SEMI-CONDUCTOR
        GAMMA RAY DETECTORS. Nucl. Instrum. Methods, 34 (May 1965),
        250-2.   Also report BNL-8692

        Cryostats for lithium-drifted germanium gamma detectors were
    designed and constructed using routine machine shop practices.
    These holders feature liquid nitrogen replenishment cycles of
    up to three weeks and activated charcoal sorption pumping which
    eliminates the need for external vacuum pumps. Low rates of
    liquid nitrogen consumption were obtained by using thin walled
    stainless steel components, in heat leakage paths, by utilizing
    commercially available super-insulated Dewars and by maintain-
    ing a sufficiently high vacuum.

147. Chasman, C.; Jones, K. W., and Ristinen, R. A.   FAST NEUTRON
        BOMBARDMENT OF A LITHIUM-DRIFTED GERMANIUM GAMMA-RAY DETECTOR.
        Nucl. Instrum. Methods, 37 (Nov. 1965), 1-8.
                                            Also report BNL-9157

        Pulse height spectra from a lithium-drifted germanium gamma
    ray detector bombarded with monoenergetic neutrons are measured
    at neutron energies of 1.2, 2.2, 4.7, and 16.3 MeV. Gamma rays
    from inelastic neutron scattering in the counter and the counter
    cryostat are identified. An E0 conversion electron line from
    the decay of the 691 keV first excited state in ^{72}Ge is very
    prominent, and at neutron energies of 4.7 and 16.3 MeV charged
    particle groups from neutron-induced reactions in the germanium
    appear. A discussion of the application of germanium gamma de-
    tectors to the study of (n, n'γ) reactions is presented. The
    detector is irradiated with a total fast neutron flux of approx-
    imately 10^{11} neutrons/cm^2 and shows no deterioration of resolu-
    tion.

148. Chaudhry, Ramesh. COMPENSATION AND PURIFICATION OF SI AND GE.
        Bhabha Atomic Research Centre, Bombay, (India). 1969. 24p.
                                                    BARC-443

        Present state of affairs as regards the purification and
    compensation of Si and Ge is reviewed from the point of view of
    fabrication of semiconductor detectors. Salient features of
    some processes of fabrication of detectors are enumerated. The
    review can be easily generalized to other processes and semi-
    conductor materials for fabrication of semiconductor detectors.

149. Chevillon-Pitollat, Pierre-Louis and Pannetier, Raymond. SPEC-
     TROMETRIC CHARACTERISTICS OF THE GAMMA EMISSION OF FISSION
     PRODUCTS: ANALYSIS OF IRRADIATED URANIUM WITH A 'GERMANIUM-
     LITHIUM' DETECTOR. (CHARACTERISTIQUES SPECTROMETRIQUES DE
     L'EMISSION GAMMA DES PRODUITS DE FISSION: ANALYSE AVEC UN
     DETECTEUR 'GERMANIUM-LITHIUM' D'UN URANIUM IRRADIE.) Commis-
     sariat a l'Energie Atomique, Saclay, (France). Centre d'Etudes
     Nucleaires. March 1969. 42p. (In French)   CEA-R-3777

     In order to study the gamma ray spectra of atmospheric aero-
     sol samples obtained with Ge-Li detectors, the spectromatic char-
     acteristics of fission product gamma-ray emission had first to
     be specified by analyzing the evolution with time of the spectra
     of irradiated uranium.

150. Christie, D. R.; Daniels, J. M., and Felsteiner, J. THE USE OF
     A LITHIUM DRIFTED GERMANIUM DETECTOR TO OBSERVE THE DE-EXCIT-
     ATION GAMMA RAYS PRODUCED BY THE INELASTIC SCATTERING OF
     2.9 MeV NEUTRONS BY COBALT AND COPPER. Nucl. Instrum. Methods,
     37 (1965), 165.

     Report discussing the use of Ge(Li) for observing x-ray from
     nuclei excited by the inelastic scattering of neutrons.

151. Chu-hsian, Wang. NUCLEAR DETECTORS AND THEIR USES. Wright-Pat-
     terson AFB, Ohio. Foreign Technology Division. Sept. 19,
     1966. 771p. Translation.   TT-67-60201; FTD-TT-65-552

     A comprehensive and systematic introduction to detection tech-
     nique in nuclear physics. It consists of four parts: the funda-
     mentals of nucleonic experiment; the various types of detectors
     (the major part of the book); synchronous application of detec-
     tors; and the reduction of experimental data.

152. Chung, Moon K. HIGH RESOLUTION GAMMA-RAY SPECTROSCOPY WITH LITH-
     IUM ION DRIFTED GERMANIUM DETECTORS, PART 1. DETECTOR FABRI-
     CATION AND PERFORMANCES. J. Nucl. Sc. (Seoul), 7 (1967),
     1-11.

     After a brief introduction to the lithium ion drift method,
     detailed fabrication procedures for planar lithium-drifted Ge
     detectors are given. Among the methods tested for ohmic con-
     tacts to the diodes, Hg-In amalgam seems to be excellent and the
     easiest way for this purpose.

153. Chung, Moon K. HIGH RESOLUTION GAMMA-RAY SPECTROSCOPY WITH LITH-

IUM ION DRIFTED GERMANIUM DETECTORS, PART 2.   APPLICATION
TO NUCLEAR PHYSICS. J. Nucl. Sc. (Seoul), 7 (1967), 12-28.

Planar-type LID Ge diodes were applied to the gamma-ray measurements of short lived nuclei with half-life less than a few
minutes, which include decays of ^{183m}W, ^{185m}W, ^{108}Ag, ^{110}Ag,
^{75m}Ge, ^{77m}Ge, ^{71}Zn, ^{161}Gd, and ^{46m}Sc. After determination of
detection efficiency for a LID Ge diode with 1 mm depletion depth,
gamma rays from ^{152}Eu, ^{154}Eu decays were re-examined, intensities are given. For routine determination of uranium isotopic
contents, gamma-rays from thermal neutron capture in ^{149}Sm and
the results of the measurements are presented. Also, pair and
Compton spectrometers employing LID Ge diodes are discussed.
The energy resolution was improved with a pair spectrometer from
30 keV FWHM to 19 keV FWHM at the double escape peak of ^{24}Na
2754 keV gamma rays.

154. Chung. M. K.; Suh, D. W.; Cho, S. W., and Kim, E. K.   SEMI-EMPIR-
     ICAL EFFICIENCY CURVE FOR A THIN GE(LI) DETECTOR AT LOW
     ENERGIES. J. Nucl. Sc. (Seoul), 10, no. 1, pt. 1 (June 1970),
     7-15. (In Korean)

155. Chwaszczewska, Janina; Dakowski, Miroslaw; Przyborski, Wincenty,
     and Sowinski, Mieczyslaw. LITHIUM-DRIFTED DETECTORS FOR
     BETA- AND GAMMA-SPECTROMETRY. Nukleonika, 10 (1965), 251-4.

The performance of lithium-drifted β and gamma-semiconductor
spectrometers is described. Lithium-drifted silicon diodes prepared on the basis of the Pell Method made possible the construction of - particle detectors with thin window and very good
characteristics. The depletion layer thickness is 2.8 mm and
active area 160 mm^2. For gamma spectroscopy, the lithium-drifting process was carried out on p-type gallium-doped germanium
samples of 5 nsec/cm resistivity and about 100 μsec lifetime
minority carriers. A compensated region 1.5 mm thick was obtained. Beta and gamma spectra for ^{207}Bi are illustrated.

156. Chwaszczewska, Janina; Dakowski, M.; Przyborski, N.; Sowinski, M.,
     and Stegner, A. SEMI-CONDUCTOR SPECTROMETERS OF PENETRATING
     NUCLEAR RADIATIONS DEVELOPED IN THE INSTITUTE OF NUCLEAR RE-
     SEARCH. Polish Academy of Sciences, Warsaw Inst. of Nuclear
     Research. Dec. 1965. 21p.   INR-681/1A,11/PS

Semiconductor spectrometers of penetrating nuclear radiation
developed in recent years in the Institute of Nuclear Research
is reviewed. The review includes guard-ring surface-barrier detectors, silicon and germanium lithium drifted detectors and

windowless silicon surface barrier-lithium drifted detectors.
The trends of development of the methods of semiconductor spectrometry are pointed out.

157. Cipolla, F. and Taroni, A. A CHANNELING EFFECT IN SOLID STATE
        DETECTORS. Nucl. Instrum. Methods, 54 (1967), 331-32.

    A channeling effect has been observed in silicon detectors
with particles. The α particles are sent on the back side of a
dEdx detector biased at a voltage such that the space charge
width is about 50 μm shorter than the detector thickness; in
this manner the non-channeled particles are stopped in the undepleted region while the channeled particles reach the space
charge region; the spectrum obtained with a multichannel analyzer shows a two peak structure.

158. Cline, J. E. ENERGY DEPENDENCE OF GERMANIUM(LITHIUM) DETECTOR
        EFFICIENCY. pp. 241-6 of "Semiconductor Nuclear Particle
        Detectors and Circuits." Brown, W. L., ed. Washington, D.C.,
        National Academy of Sciences, 1969.
                                        Also report CONF-670520

    From Conference on Semiconductor Nuclear Particle Detectors
and Circuits, Gatlinburg, Tenn.

    Efficiency measurements were made for six Ge(Li) detectors
of various sizes and types using gamma rays of 50 keV to 10 MeV
in order to determine the energy dependence of efficiency.

159. Cline, J. E. and Heath, R. L. GAMMA-RAY SPECTROMETRY OF NEUTRON-
        DEFICIENT ISOTOPES. Idaho Nuclear Corp., Idaho Falls. Annual
        Progress Report. May 1969. 84p.   IN-1130

    The primary objective of this program was to produce neutron-
deficient nuclides to obtain high-resolution gamma-ray spectra
and basic information on Gamma-ray Spectrum Catalogue. The emphasis in this effort is to provide nuclear laboratories and ex-
perimenters with these data obtained with state-of-the-art gamma-
ray spectrometers. To make this information most useful, standard methods for obtaining and analyzing data obtained with Ge(Li)
spectrometers are being developed. Neutron-deficient nuclides
were produced primarily by proton bombardment of target materials using (p, xn) reactions and by (n,p) and (n,α) reactions
using fast neutrons. Spectra were obtained for approximately
30 nuclides. Data on these nuclides precise energies and intensities for the primary gamma rays. Considerable effort was also
made in the study of the operating characteristics of standard

laboratory spectrometers. Systematic studies were made of the detection efficiency of Ge(Li) detectors for gamma rays up to 10 MeV.

160. Cline, J. E. STUDIES OF DETECTION EFFICIENCIES AND OPERATIONS CHARACTERISTICS OF GE(LI) DETECTORS. IEEE Trans. Nucl. Sci., NS-15, no. 3 (June 1968), 198-213.

161. Coche, A. SEMICONDUCTOR DETECTORS: EVOLUTION PERFORMANCE FUTURE OUTLOOK. Nucleus (Paris), 9 (1968), 114-21. (In French)

After reviewing the various types of detectors and the performance they actually attain, the problems presented in the realization of use of these counters are set forth. Such problems discussed are the detection of heavy particles using n-p junction surface, barrier counters, and the detection of gamma rays using Ge(Li) counters. The future prospects of their development are also considered.

162. Coche, A. SEMI-CONDUCTOR DETECTORS OF NUCLEAR RADIATION. Ind. At., 9 (1965), 61-9. (In French)

The characteristics of diodes with a n-p linkage and with surface barriers and their uses are examined. A method for compensating a material of the p-type by the migration of lithium is also presented.

163. Coche, A. SEMICONDUCTOR NUCLEAR RADIATION DETECTOR. Onde Elect., 49 (July-Aug. 1969), 791-5. (In French)

After a review of the general characteristics of semiconductor counters, the main problems encountered in the applications to radiation detection and nuclear spectroscopy are examined.

164. Coleman, J. A. GERMANIUM FOR GAMMA-RAY DETECTORS: A REVIEW OF CURRENT PROBLEMS. pp. 37-41 of "Lithium-Drifted Germanium Detectors. Proceedings of a Panel on the Use of Lithium-Drifted Germanium Gamma-Ray Detectors for Research in Nuclear Physics, Vienna, 6-10, June 1966." Vienna International Atomic Energy, 1966.

The performance of germanium gamma radiation detectors depends, to a large extent, on the quality of the germanium used to produce the devices. At present, the demand for better resolution, faster response times, and larger volume detectors imposes severe requirements on the germanium for detectors in terms

of crystal purity and perfection. The specifications commonly
used to describe germanium for detectors, such as resistivity,
carrier lifetime, and etch pit density, were originally derived
from those used by the semiconductor device industry. The rele-
vance of these specifications for material to produce a detector
with good performance, including long-term stability, is not
well understood. A semiconductor detector requires extreme perfection throughout a large-volume single crystal. Such is not
the case for common transistors and diodes which operate at
much higher currents and noise levels and with much smaller
volumes.

165. COMMERCIALLY AVAILABLE SEMICONDUCTOR DETECTORS AND PREAMPLIFIERS.
Nucleonics, 22, no. 5 (May 1964), 62-7.

Tables of properties of commercial semiconductor detectors--
silicon detectors: surface-barrier, phosphorus-diffused, total-
ly depleted, and lithium-drifted; germanium lithium-drifted de-
tectors--and preamplifiers are given.

166. CONFERENCE ON SEMICONDUCTOR NUCLEAR PARTICLE DETECTORS AND CIR-
CUITS. Proceedings of a Conference, Gatlinburg, Tenn.
May 15-18, 1967. Washington, D. C., National Academy of
Sciences, 1969.   CONF-670520

167. CONTROL OF LITHIUM MIGRATION AND PRECIPITATION IN GERMANIUM SIN-
GLE CRYSTALS. pp. 176-81 of "Nucleonics in Aerospace."
Polishuk, Paul, ed. New York, Plenum Press, 1968.
Also report CONF-670714

From 2nd International Symposium on Nucleonics in Aerospace,
Columbus, Ohio.

Lithium mobility alone is not a sufficient criterion to char-
acterize germanium single crystals for suitability in making radiation detectors. The second important drifting is described
which allows the extraction of meaningful lithium precipitation
data.

168. Cooper, J. A.; Perkins, R. W., and Kosorok, J. R. ANTICOINCIDENCE-
SHIELDED DUAL GE(LI) GAMMA-RAY SPECTROMETER FOR LOW-LEVEL
COUNTING. Trans. Amer. Nucl. Soc., 13 (June-July 1970),
76-7.   Also report CONF-700608

169. Cooper, J. A.; Rancitelli, L. A.; Perkins, R. W.; Haller, W. A., and Jackson, A. L. AN ANTICOINCIDENCE-SHIELDED GE(LI) GAMMA-RAY SPECTROMETER AND ITS APPLICATION TO NEUTRON ACTIVATION ANALYSIS. Battelle-Northwest, Richland, Wash. Pacific Northwest Lab. Aug. 2, 1968. 15p.   BNWL-SA-2009

From International Conf. on Modern Trends in Activation Analysis, Gaithersburg, Md.

The design and performance of an anticoincidence-shielded lithium-drifted germanium gamma spectrometer with high efficiency, resolution, and Compton reduction are presented. The use of the spectrometer in nondestructive activation analysis is also reviewed.

170. Cooper, J. A.; Rancitelli, L. A.; Perkins, R. W.; Haller, W. A., and Jackson, A. L. AN ANTICOINCIDENCE SHIELDED GE(LI) GAMMA-RAY SPECTROMETER AND ITS APPLICATION TO NEUTRON ACTIVATION ANALYSIS. pp. 1054-61 of the "Modern Trends in Activation Analysis." Vol. 2. Washington, D. C., National Bureau of Standards, 1969. 1969.   CONF-681003-v.2

An anticoincidence shielded Ge(Li) gamma-ray spectrometer, which was designed to provide a high degree of Compton reduction along with high efficiency and resolution, is described and its use in extending the applications of nondestructive activation analysis is illustrated.

171. Cooper, J. A.; Wogman, N. A., and Perkins, R. W. AN ANTICOINCIDENCE-SHIELDED GE(LI) GAMMA-RAY SPECTROMETER FOR HIGH SENSITIVITY COUNTING. Battelle-Northwest, Richland, Wash., Pacific Northwest Lab. March 1968. 12p.
                    BNWL-SA-1522; CONF-680207-5

An anticoincidence-shielded Ge(Li) gamma-ray spectrometer for high sensitivity counting is described. The system consists of a 20 cm coaxial diode centered inside a 26" diameter by 24" thick plastic phosphor. The source is placed inside the shield and both the coincidence and anticoincidence spectra are recorded simultaneously. The anticoincidence shield reduces the ^{137}Cs Compton edge by a factor of 7.5 to provide a peak to Compton edge ratio of 150:1 without reducing the photopeak efficiency by more than 3%. This system also provides a two to four-fold lower natural radio-nuclide background that similar sized diodes.

172. Cooper, J. A.; Wogman, N. A.; Perkins, R. W., and Rancitelli, R.A. THE APPLICATION OF ANTICOINCIDENCE-SHIELDED GE(LI) SPECTRO-

METERS TO RADIOCHEMICAL PROBLEMS. Trans. Amer. Nucl. Sci.,
12 (June 1969), 65-7.   Also report CONF-690609

From 15th Annual Meeting of the American Nuclear Society,
Seattle, Washington.

173. Cooper, J. A. and Jackson, A. L. THE EFFICIENCY CALIBRATION OF
     A LARGE-VOLUME GE(LI) GAMMA-RAY SPECTROMETER. pp. 118-21
     of "Pacific Northwest Laboratory Annual Report for 1968."
     Vol. 2.   USAEC Division of Biology and Medicine. Battelle-
     Northwest, Pacific Northwest Lab., Richland, Washington.
     1969.   BNWL-1051-pt.2

     The relative and absolute efficiency functions for a large
     volume Ge(Li) gamma-ray spectrometer reported for a point source
     at various distances from the detector.

174. Cooper, J. A. EVALUATION OF GE(LI) COMPTON SUPPRESSION SPECTRO-
     METERS FOR RADIOCHEMICAL ANALYSIS. pp. 121-7 of "Pacific
     Northwest Laboratory Annual Report for 1968." Vol. 2.   USAEC
     Division of Biology and Medicine. Battelle-Northwest, Paci-
     fic Northwest, Richland, Wash. 1969.

                                               BNWL-1051-pt.2

175. Cooper, J. A. FACTORS DETERMINING THE ULTIMATE DETECTION SENSI-
     TIVITY OF GE(LI) GAMMA-RAY SPECTROMETERS. Nucl. Instrum.
     Methods, 82 (1970), 273-7.   Also report CONF-700608

     An equation is derived which relates the peak detection sen-
     sitivity of Ge(Li) gamma-ray spectrometers to their efficiency,
     resolution and background interference. A figure-of-merit is
     derived which is directly related to the detectors sensitivity
     and a procedure is suggested for comparing the quality of vari-
     ous Ge(Li) spectrometers.

176. Cooper, J. A. and Browell, G. L. A LARGE COAXIAL GE(LI) DETEC-
     TOR WITH A PLASTIC ANTICOINCIDENCE SCINTILLATION FOR ACTI-
     VATION ANALYSIS. Nucl. Instrum. Methods, 51 (1967), 72-6.

     A gamma ray spectrometer has been constructed which makes
     use of a Ge(Li) detector with 35 cm^3 active volumes surrounded
     by a plastic scintillator annulus for Compton suppression.
     Efficiencies between 0.1 and 1 percent are obtained with reso-
     lutions between 6 and 7 keV. Compton suppression of about a
     factor of two is obtained.

177. Currie, R. L.; McPherson, R., and Morrison, G. H. A COINCIDENCE-
     ANTICOINCIDENCE SYSTEM FOR ACTIVATION ANALYSIS EMPLOYING A
     SPLIT NAI(TL) ANNULUS AND A LARGE VOLUME GE(LI) DETECTOR.
     pp. 1062-8 of "Modern Trends in Activation Analysis." Vol. 2.
     DeVoe, James R., ed. Washington, D. C., National Bureau of
     Standards, 1969.   CONF-681003-v.2

     From International Conference on Modern Trends in Activation
     Analysis, Gaithersburg, Md.

     A system was devised to permit several simultaneous experi-
     ments to be performed using a detector combination with a single
     multichannel analyzer. A large Ge(Li) detector with very good
     resolution and high peak-to-Compton ratio was inserted midway
     into the 3-in.-dia. tunnel of an 8-in.-dia. by 10-in.-long NaI(Tl)
     annulus. The source was located on the cryostat window for max-
     imum sensitivity. A block diagram of the electronics of the sys-
     tem is shown. The application of this system to activation analy-
     sis of complex systems is described.

178. Dakovski, M.; Pshiborski, V.; Sventchak, M.; Sowinski, M., and
     Khvashchevska, Ya. POLUCHENIE GERMANIEVYKH SPEKTROMETROV
     γ IZLUCHENIYA PO METODIE DREIFA IONOV LITIYA. (PREPARATION
     OF GERMANIUM γ SPECTROMETERS BY LITHIUM DRIFT METHOD). In-
     stitute Nuclear Research, Warsaw, (Poland). Dec. 1964. 13p.
                                                   INR-592/11/P1

     Germanium diodes for gamma ray spectrometry using the lithi-
     um-ion drift technique and p-type germanium have been fabricated.
     Active depths up to 1.5 mm were obtained. The spectrum source
     radioactive elements using these detectors (working at 90°K) are
     presented. The resolution for ^{181}Hf (gamma-energy of 615 keV)
     is 7.2 keV for ^{60}Co (gamma-line of 1333 keV) is 12 keV.

179. Dau, Gary J. SOLID-STATE GAMMA SPECTROMETRY DEVELOPMENTS.
     pp. 4-10 of "Northwest Pacific Lab. Annual Report, 1965.
     Vol. 4. Instrumentation." May 1966.   BNWL-4

     Fabrication and testing of Li-drifted germanium radiation
     detectors are described. The detectors were fabricated to de-
     termine their characteristics and to explore their potential
     uses. Data are included on detector efficiency and gamma energy
     spectra of ^{239}Pu and ^{241}Am.

180. Davies, D. E. and Webb, P. P. SURFACE POTENTIAL MEASUREMENTS OF
     LITHIUM DRIFTED GERMANIUM DIODES. n.d. 15p.

From IEEE Nucl. Sci. Symposium, San Francisco.

A potential probe was built to measure the distribution of potential on the etched surfaces of lithium-drifted Ge diodes. A vibrating condenser probe was used. Initial measurements with the probe confirmed a theoretical model proposed by McIntyre to explain previously observed unusual behavior of some lithium-drifted diodes. The probe was also used to determine the effect of various surface treatments designed to produce Ge diodes having high breakdown voltages and low leakage currents.

181. Davydov, E. F.; Sukhikh, A. V.; Syuzev, A. V., and Ivanov, V. I. APPLICATION OF SEMICONDUCTOR GAMMA-SPECTROMETER FOR STUDYING SOME CHARACTERISTICS OF CERAMIC FUEL ELEMENTS. **At. Energ. (USSR)**, 27 (Sept. 1969), 224-5. (In Russian)

Descriptions are given for a design and performances of a Ge(Li)-semiconductor gamma-spectrometer with a sensitivity thickness of 8 mm·and resolving power along the ^{137}Cs line of 6 keV. A collimating device is capable of varying the gamma-quantum flux from irradiated fuel elements where the geometric dimensions of the collimating opening varied from $5 < \phi < 20$ mm and the length of collimator was 1500 mm. Physical characteristics of uranium dioxide in a stainless steel cladding were studied at 500-600°C.

182. Day, Robert B. and Palms, John M. HIGH RESOLUTION MEASUREMENTS OF GAMMA RAYS FROM NEUTRON INELASTIC SCATTERING. Los Alamos Scientific Lab., N. Mex. 1964. 5p.
                          LA-DC-7096; also report CONF-650706-20

A comparison of the new lithium-drifted germanium and the NaI neutron detector is presented.

183. Deal, R. A.; Gettings, J. F., and James, D. B. PLUTONIUM WASTE ASSAY UTILIZING A LARGE VOLUME GE(LI) DETECTOR. pp. 61-76 of "Proceedings of the Tenth Annual Meeting of the Institute for Nuclear Materials Management, Las Vegas, Nevada. April 28-30, 1969." Columbus, Ohio, Institute of Nuclear Materials Management. 1969.   CONF-690411

The results of a limited study of the application of Ge(Li) gamma spectroscopy to the assay of PuO_2-contaminated waste are presented. The method is based on the use of commercially available instrumentation components, and requires that waste packages be standardized within the plant. More important than establishing an assay method for PuO_2 in polyviny-chloride (PVC)

waste, these results have been presented to highlight the feasibility of this approach to the assay of many forms of waste materials contaminated with many types of nuclear materials. By
utilizing the other specific gamma activities and preparing appropriate standard curves, the technique can be applied to other
forms of low-density waste, contaminated with metal or compounds
of plutonium and uranium and their mixtures, with varying isotopic compositions.

184. Dearnaley, G. and Hardacre, A. G. FURTHER NOTE ON ION IMPLANTED
     GERMANIUM DETECTORS. Nucl. Instrum. Methods, 82 (1970),
     187-8.

185. Dearnaley, G.; Hardacre, A. G., and Rogers, B. D. ION IMPLANTED
     GERMANIUM PARTICLE DETECTORS. Nucl. Instrum. Methods, 71
     (1969), 86-92.   Also report EUR-4269

     Thin-windowed germanium detectors have been prepared by ion
     implantation with gallium followed by a conventional lithium
     ion-drift. Such a detector has shown an overall energy resolu-
     tion of 37 keV (FWHM) for 34 MeV protons. Of this total line-
     width, the contribution due to the detection system alone is es-
     timated to be 22 keV. A description is given of a cryostat as-
     sembly which allows a dE/dx and E particle discrimination by
     means of a silicon and germanium detector telescope. Other
     applications of thin-windowed germanium detectors are also con-
     sidered.

186. Dearnaley, G.; Gibbons, P. E., and Ellis, R. LARGE DIAMETER
     GERMANIUM CRYSTALS FOR GAMMA-RAY SPECTROSCOPY. Trans. Nucl.
     Sci., NS-17, no. 3 (June 1970), 282-6.
                                 Also report CONF-700301

     From 12th Scintillation and Semiconductor Counter Symposium,
     Washington, D. C.

     The progress made in the development of large diameter ger-
     manium crystals for gamma-ray spectrometry is reviewed. These
     are vacuum pulled and have been shown to be consistently free
     of oxygen. Planar devices of 40 sq. cm. area and depletion
     thickness of 8 mm have given an energy resolution better than
     3.7 keV at 1.33 MeV. The relationship between detector perform-
     ance and crystal properties is examined for possible correlation.

187. Dearnaley, G. and Northrup, D. C. SEMICONDUCTOR COUNTERS FOR
     NUCLEAR RADIATIONS. 2nd ed. London, F. N. Spon. 1966. 45p.

A review of recent information on fabrication and application of large lithium-drifted germanium counters, on processes
of surface barrier formation, on applications in fields other
than nuclear physics and on developments in ancilliary electronics.

188. Dearnaley, G. SOLID-STATE RADIATION DETECTORS. Contemp. Phys.,
       8 (Nov. 1967), 607-26.

   A historical survey of the development of solid-state detectors is given, and it is shown why semiconductor detectors are
superior to the earlier crystal counters. The physical processes which occur during the detection of nuclear radiation in
a solid-state device are considered in detail, and the merits
of the reverse-biased semiconductor junction in silicon or germanium are stated. Factors which determine the energy resolution of such a detector are analyzed, and also the effects of
radiation damage. The preparation of such detectors is not
treated in detail but the physical principles on which the important types of detector depend are described. The applications
of solid-state detectors in nuclear physics, radiochemical analysis, space research, medicine and biology are surveyed.

189. DeBruin, M. and Hoekstra, W. PREPARATION OF A LOW CAPACITANCE
       GE(LI) DETECTOR. Nucl. Instrum. Methods, 51 (1967), 353-55.

   The preparation of a low capacitance Ge(Li) detector with
a thick depletion layer, using a modified coaxial drifting technique is described. The properties of the detector are demonstrated.

190. de Castro, Faria N. V. and Levesque, R. J. A. A MONTE-CARLO
       CALCULATION OF INTRINSIC DOUBLE-ESCAPE PEAK EFFICIENCY OF
       CYLINDRICAL LITHIUM-DRIFT GERMANIUM GAMMA RAY DETECTORS.
       IEEE Trans. Nucl. Sci., NS-13, no. 3 (June 1966), 363-66.

   Calculations of the intrinsic double-escape peak efficiency
of cylindrical Li-Ge detectors for parallel beam of gamma rays
ranging in energy from 1.5 to 10 MeV were carried out using a
computer code based on the Monte Carlo method. The detectors
considered are 18 mm in diameter with 2, 3.5, 4, 5, 8, and 10 mm
depletion depth.

191. de Castro, Faria N. V. and Levesque, R. J. A. PHOTOPEAK AND
       DOUBLE-ESCAPE PEAK EFFICIENCIES OF GERMANIUM LITHIUM DRIFT
       DETECTORS. Nucl. Instrum. Methods, 46 (1967), 325-332.

This paper presents theoretical calculations by the Monte Carlo method of the intrinsic efficiencies, photo and double-escape fractions of Ge(Li) drift detectors for parallel beams of gamma rays in the range of energy from 100 keV to 10 MeV. For the photopeak efficiencies the calculations were made from 100 keV to 1.5 MeV and for the double-escape peak efficiencies, they were carried for energies greater than 1.5 MeV. The detectors considered were cylinders having 18 mm dia. and depletion depths of 2, 3.5, 4, 5, 6, 7, 8, 9, and 10 mm. The results were compared with experimental values obtained by Evans et al., with 18 x 3.5 mm detectors, for both the photopeak and the double-escape peak efficiencies and with a 5 cm x 8 mm for the full energy peak efficiency.

192. De Geer, L. E. STUDIES OF MIXED FISSION PRODUCTS FROM THERMAL FISSION OF ^{235}U WITH A GE(LI) DETECTOR. April 1970. 19p.
FOA-4-4420-28

Gamma spectra from fission products of reactor-irradiated ^{235}U were reduced with a 7.2 cm^3 crystal and analyzed by a 1600-channel analyzer. The Ge(Li) detector was shown to be superior to the NaI detector for analysis of complicated spectra. A data table is presented with counting rate for every known actual photopeak. The purpose of this work was to create tools for the analysis of fallout from nuclear explosion.

193. Degtyarev, Yu. G. and Protopopov, V. N. GERMANIUM GAMMA SPECTRO-METER. Prib. Tekh. Ekep., no. 2 (Mar.-Apr. 1969), 33-7. (In French)

The procedure for compensation of acceptor impurities in Ge by drifting Li ions at 56°C is described. The design character-istics of the cryostat for the operation of the detector at li-quid nitrogen temperatures are given. The energy resolution for the gamma peak from ^{170}Tm (E_γ=84.3 keV) is 4.1 keV. The spectra from ^{170}Tm, ^{51}Cr, ^{137}Ga, ^{46}Sc and ^{60}Co were measured.

194. de Jesus, A. S. M. THE MANUFACTURE OF GE(LI) DETECTORS. Council for Scientific and Industrial Research, Pretoria, South Africa. National Physical Research Lab. 1969. 15p.
CSIR-283

Ge(Li) detectors have been manufactured at the National Phy-sical Research Laboratory using both the planar and coaxial con-figurations. The manufacturing procedure and characteristics of the detectors are given.

Problems associated with the base material and other problems encountered during the various phases of manufacture are discussed. The elimination of the need for a reetch when transferring a detector from one cryostat to another is also described.

195. De Laet, L. PRODUCTION AND FUNDAMENTAL PROPERTIES OF GERMANIUM SINGLE CRYSTALS WITH RESPECT TO THEIR APPLICATION IN GAMMA-RAY DETECTORS. Paper 2 of "The Proceedings of the Meeting on the Applications of Ge(Li) Detectors in Science, Technology, Medicine and Industry, Brussels, Oct. 20, 1967." 1968. 16p.   CONF-671078; BLG-425

Germanium refining and single crystal pulling techniques are briefly discussed. Fundamental properties of germanium single crystals considered include resistivity and type. Hall effect, minority carrier lifetime, physical perfection, oxygen determination, and migration and precipitation of lithium.

196. de Lange, P. W. and Snyman, G. C. GE(LI) DETECTORS: A REVIEW OF THEIR PROPERTIES AND THEIR APPLICATION IN GAMMA SPECTROMETRY. Atomic Energy Board. Pretoria, (So. Africa). Dec. 1966. 26p.   PEL-135

The properties of planar and coaxial Ge(Li) detectors in gamma-ray spectroscopy are reviewed. Three different cryostat systems to operate Ge(Li) detectors at low temperatures are compared. Recommendations are proposed for would-be users of this new tool in gamma-spectrometry.

197. de Lange, P. W. and Bigham, C. B. MULTI-ELEMENT DETECTORS FOR ACTIVATION RESONANCE INTEGRAL AND RESONANCE SPECTRUM MEASUREMENTS. Nucl. Appl., 4 (March 1968), 190-5.

A lithium-drift germanium gamma-ray spectrometer can resolve and measure separately the activities of six elements mixed together in multi-element detectors. The activators and main resonance energies were ^{115}In (1.4eV), ^{197}As (47eV), ^{198}Pt (96eV), ^{98}Mo (467eV), and ^{68}Zn (514eV). The detectors were made by loading the six elements into epoxy resin. The results showed that good uniformity had been obtained and gave reduced resonance integrals for ^{68}Zn 0.40±0.066 (for ^{69}Zn formation), ^{75}As (58.8± 3.46), ^{98}Mo (6.3-0.8b), and ^{198}Pt (50-6b).

198. Delor, Michel. GAMMA-RAY SPECTRA COMPUTER CODE. PART 1. Commissariat a l'Energie Atomique, Cadarache, (France). Centre d'Etudes Nucleaires. Sept. 1970. 46p.
                                             CEA-N-1381 (part 1)

A computer code for automatic analysis of complex gamma ray spectra from multichannel analyzer and a Ge(Li) detector is presented. The method enables the separation of close peaks: doublets, triplets, etc., At the moment this program is limited to calculate the energies and the areas of the significant peaks.

199. Delporte, R. GERMANIUM DETECTORS COMPENSATED BY LITHIUM ION MIGRATION. Part 1. CHARACTERISTICS AND ACTUAL PERFORMANCE IN GAMMA SPECTROMETRY. Ind. Chim.Belge, 33 (1968), 335-42. (In French)

Resolution, linearity, and efficiency of the Ge(Li) detector are described. The principal sources of noise which limit resolution are noise of electronic origin, fluctuations in the number of electron-hole pairs created, and fluctuations in the yield of collected charge carriers. Noise level contributions are calculated for a high-performance spectrometer using a low-noise preamplifier provided with a FET. The linearity between 0 and 2600 keV has been demonstrated. The variation of efficiency with photon energy is a result of photoelectric and Compton effects and pair production. The efficiency above 200 keV is always superior to that calculated from photoelectric absorption considerations. The efficiency of a 54 cm^3 Ge(Li) diode is comparable to that of a 1 x 1-1/2 in. NaI(Tl) crystal. For photons of more than 1.5 MeV energy, experiment deviates badly from theory.

200. Delporte, R. GERMANIUM DETECTORS COMPENSATED BY LITHIUM ION MIGRATION. Part 2. APPLICATIONS TO ELEMENTAL CHEMICAL ANALYSIS. Ind. Chim. Belge, 33 (1968), 440-8. (In French)

A survey of recent work illustrating the utility of the high resolution of Ge(Li) detectors for elemental analysis is given; in particular, examples indicating possibilities for improved neutron activation analysis are chosen. The resolution of the $K\alpha_1$, $K\alpha_2$, $K\beta_1$, and $K\beta_2$ of heavier elements and mixtures of them is achieved.

201. Demuynck, J. and Uyttenhove, J. ROUTING UNIT FOR MEASUREMENTS ON SHORT-LIVING ISOMERS. Nucl. Instrum. Methods, 74 (1969), 97-100.

An apparatus for investigations on short-living isomers in the region of msec and μ sec is described. By measurement of a gamma-ray spectrum in consecutive time intervals, a maximum of information on energy and time is obtained.

202. Den Boer, J. A. and Hofker, W. K. SPECIAL SEMICONDUCTOR DETEC-
      TORS FOR GAMMA RAY AND PARTICLE SPECTROSCOPY. pp. 116-125
      of "Gemeinsame Tagung Ueber Halbleiterdetektoren in Stra-
      hlenschutz und Strahtenmesstechnik... 1967." (Joint Meeting
      on Semiconductor Detectors in Radiation Protection and Radi-
      ation Measurement Techniques... 1967). 1967. CONF-670545

      A survey is given of our work on alpha-particle, Beta-par-
      ticle, and gamma ray detectors, detectors for biological and
      in vivo measurements and detectors for dE/dx x E telescope sys-
      tems. More in detail the checker-board detector, a totally
      depleted semiconductor detector with position indication in two
      dimensions will be discussed.

203. Deshpande, R. Y. DETECTION MECHANISM IN SEMICONDUCTOR COUNTERS.
      Nucl. Instrum. Methods, 57 (1967), 125-130.

      A review of the detection mechanism in junction type semi-
      conductor counters. Discusses the processes of energy loss, the
      creation of charge pairs, signal formation in the presence of
      background noise, and charge collection and separation.

204. Dessert, Robert Allen. THE CONSTRUCTION OF A LITHIUM DRIFTED
      GERMANIUM SOLID STATE DETECTOR SYSTEM WITH SPECIAL ATTENTION
      TO ^{105}AG AND ^{192}IR. Ohio State University, Columbus. 1965.
      72p. Thesis   Also report AD-620434

      A relatively inexpensive cryostat was constructed for opera-
      tion of Li drifted Ge gamma detectors at pressures below
      0.0000001 mm of Hg and at 77°K. Fundamental data were obtained
      on a Li-drifted Ge detector, including relative counting effi-
      ciency, resolution and peak-to-total ratio. The spectra of
      ^{192}Ir and ^{105}Ag were investigated using the solid-state detec-
      tor. A technique of segmentation and expansion of energy sub-
      groups with a multichannel analyzer is proposed as a means to
      provide more points in a given peak of spectrum taken on the
      detector.

205. DETECTORS OF IONIZING PARTICLES AND RADIATION. pp. 44-52 of
      "Oak Ridge National Lab. Instrumentation and Controls Divi-
      sion Annual Progress Report for Period Ending Sept. 1, 1965."

      A report is presented on detectors of ionizing particles and
      radiation. Topics discussed are production and encapsulation of
      lithium-drifted germanium detectors, basic instrumentation in
      radioisotope applications, alpha scintillation detector response
      in gamma ray and neutron backgrounds, ion-chamber current fluc-

tuations produced by neutron and gamma sources, PCP 3 ionization
chambers, spherical BF_3, proportional counters, and radiation
damage in cadmium sulfide and cadmium telluride.

206. DETECTORS OF IONIZING PARTICLES AND RADIATION. pp. 15-22 of
     "Instrumentation and Controls Division Annual Progress Re-
     port for Period Ending Sept. 1, 1969." Sadowski, G. S., ed.
     Oak Ridge National Lab., Tenn. 1969.   ORNL-4459

     Includes the development of germanium semiconductor detec-
     tors, and other proportional counters.

207. Dewit, R. C. and McKenzie, J. M. RECTIFIED AC DRIFTING OF GER-
     MANIUM DETECTORS. Nucl. Instrum. Methods, 83 (1970), 142-4.

     The technique of drifting the lithium in a rectified ac
     field during fabrication of lithium drifted detectors is de-
     scribed. The resultant time dependent current waveforms when
     interpreted are used to monitor the drift.

208. Dewit, R. C. and McKenzie, J. M. SURFACE PREPARATION TO OBTAIN
     GOOD I-V CHARACTERISTICS ON GERMANIUM LITHIUM DIODES. IEEE
     Trans. Nucl. Sci., NS-15 (June 1968), 352-58.

     High voltage breakdown low current diodes have been made by
     the simple and reproducible process of quenching the final
     H_2O_2-HF etch of the diode with an acqueous salt solution. Of
     these salts tried, $CaCl_2$ gives the best overall performance.
     Planar diodes 28 cm diameter and 5 mm drift depth have been made
     with leakage currents $<5x10^{-10}A$ for fields up to 600 V/mm.

     The diode surface noise on these units was less than 0.3 keV.
     The surface treatment also gives some surface passivation.

209. Dittner, A.; Hartmann, G., and Klein, J. W. ZERO CROSS TRIGGER
     WITH RISE TIME CORRECTION FOR TIMING OF GE(LI) DETECTOR
     PULSES. Nucl. Instrum. Methods, 89 (1970), 73-5.

     A method is proposed by which the timing signal from a Ge(Li)
     detector may be improved by a factor of 8 compared with leading
     edge triggering. The detector pulse in the timing channel is
     limited and double delay-line clipped, so that the charge of the
     pulse between start and zero cross depends only on the slope of
     the pulse. This charge is integrated on a capacitor. At zero
     cross time, a constant current discharge of the capacitor is
     started, and a time correction is thus added to the zero cross

signal, depending on the charge of the first part of the pulse
corresponding to a rise time correction. The method is exact
for pulses with linear slope and the maximal error is estimated
for real Ge(Li) pulse form.

210. Dolev, A.; Adam, G., and Katriel, J. A NOTE ON THE WALFORD-DOUST
     METHOD FOR RAPID CALIBRATION OF GERMANIUM SPECTROMETERS.
     Nucl. Instrum. Methods, 68 (Feb. 1, 1969), 176.

211. Donnelly, D. P. and Wiedendeck, M. L. THE CALIBRATION OF A GE(LI)
     GAMMA-RAY SPECTROMETER FOR ENERGY AND RELATIVE INTENSITY MEA-
     SUREMENTS. Nucl. Instrum. Methods, 57, no. 2 (1967), 219-26.

     Two aspects of Ge(Li) gamma-ray spectrometry are discussed.
     The non-linearity of an amplifier-analyzer system was measured
     using a precision pulser and was checked with gamma-ray standards.
     The overall accuracy was 2 parts in 10^4. The relative detection
     efficiency of gamma-ray emission rates. Two Ge(Li) detectors
     with active volumes of 3 cm^2 x 0.5 cm^2 x 0.5 cm were calibrated
     in the energy range 80-3200 keV.

212. Donnelly, D. P. and Wiedendeck, M. L. THE RELATIVE DETECTION
     EFFICIENCY CALIBRATION OF A GE(LI) DETECTOR AT LOW ENERGIES.
     Nucl. Instrum. Methods, 64 (Sept. 1968), 26-8.

     The relative detection efficiency of gamma-ray full energy
     peaks in the energy region 20-100 keV was determined using a
     modified pair-point method. An overall accuracy of 15% was ob-
     tained in this energy region.

213. Donovan, P. F. USER'S VIEW. pp. 772-5 of "Semiconductor Nuclear-
     Particle Detectors and Circuits." Brown, W. L., ed. Wash-
     ington, D. C., National Academy of Sciences, 1969.
                                        Also report CONF-670520

     The performance and capabilities of Ge(Li) detectors for
     charged particle and gamma spectroscopy are discussed from a
     user's point of view.

214. Dooley, John A. MEASUREMENT AND COMPARISON OF DETECTOR EFFICIEN-
     CIES. pp. 1043-8 of "Modern Trends in Activation Analysis."
     Vol. 2. DeVoe, James R., ed. Washington, D. C., National
     Bureau of Standards, 1969.   CONF-681003

From International Conference on Modern Trends in Activation Analysis, Gaithersburg, Md.

For comparison of efficiencies of NaI(Tl) and Ge(Li) detectors, measurements of a standard source at 2 or more distances from the detector were made, and a plot was made of the inverse square root of the count area as a function of source–detector distance. Two quantities were derived from the plot that together allow computation of efficiency for any source–detector distance including the maximum efficiency attainable by the detector. For the selection of best detector, it is recommended that the efficiency factor and the distance of closest approach for the energy region of interest should be the 2 parameters used in the comparison.

215. Douglass, Terry Dean. THE APPLICATION OF FILTERS TO TIME ANALYSIS OF SIGNALS FROM GE(LI) DETECTORS. Tennessee Univ., Knoxville. 1968. 140p. Thesis.

An optimum filter which minimizes the effect of noise on time measurement was determined and the minimum timing error associated with this optimum filter was found. For the signal and noise from a Ge(Li) detector, charge-sensitive preamplifier system, the certain realizable RC filters on the reduction of the time measurement error due to noise is presented. The filters examined were time-invariant and time-variant RC highpass, RC low-pass, and the combination of RG low-pass, RC high-pass filters. Both theoretical and experimental data were examined to determine the best filter.

216. Douglass, Terry Dean and Williams, C. W. A COMPARISON OF VARIOUS FILTER AND DISCRIMINATION TECHNIQUES ON TIMING WITH GE(LI) DETECTORS. IEEE Trans. Nucl. Sci., NS-16, no. 1 (Feb. 1969), 87-91.

This paper presents an experimental examination of the effect of several filters and two discriminator techniques on the time measurement of the signals from coaxial Ge(Li) detectors.

217. Drexler, G.; Perzl, F., and Panzer, W. MEASUREMENTS ON X-RAY SPECTRA WITH GERMANIUM DETECTORS. (MESSUNGEN VON ROENTGEN-SPEKTREN MIT GERMANIUM-DETEKTOREN.) pp. 171-76 of "Soc. Europeenne de Protec. Centre des Rayonnements Detectors in Radiation Protec. and Radiation Measurement Tech. Sept. 1967." 1967. (In German)

Measurements of x-ray spectra with germanium detectors were

compared with those obtained with a NaI scintillation detector.
The energy dependence of radiation measurement devices in the
lower energy range was determined. The results showed that Ge
detectors can be used advantageously in the low energy range and
more accurate measurements are possible in an area which is of
importance both for radiation protection and dosimetry.

218. Drexler, G. and Perzl, F. SPECTROMETRY OF LOW-ENERGY GAMMA-
         AND X-RAYS WITH GE(LI) DETECTORS. Nucl. Instrum. Methods,
         48 (1967), 332-34.

219. Drummond, W. E. HIGH PURITY GERMANIUM DETECTORS. IEEE Trans.
         Nucl. Sci., NS-18, no. 2 (April 1971), 91-100.

220. Dubois, J. THE USE OF A LITHIUM-DRIFTED GERMANIUM SOLID STATE
         DETECTOR AS A PRECISION GAMMA RAY SPECTROMETER. Ark. Fys.,
         30 (1965), 497-498A.

221. Eckerman, K. F. and Kastner, Jacob. ANALYSIS OF THE RESPONSE
         OF GE(LI) AND SI(LI) DETECTORS TO THE 46.5 keV GAMMA RAY OF
         ^{210}Pb. pp. 28-32 of the "Radiological Physics Division
         Annual Report. Radiological and Health Physics, July 1969-
         June 1970." Argonne National Lab., Ill. 1970.
                                            ANL-7760 (Part 1)

     Ge(Li) and Si(Li) detectors were evaluated as potential de-
tectors for in vivo measurements of ^{210}Pb. An analytical approach to the evaluation was developed which leads to optimization of the signal-to-background ratio and thus provides the
experimenter with information upon which to base selection of a
detection system.

222. Edwards, W. D. and Wilburn, C. OXYGEN-FREE, ALUMINUM DOPED GER-
         MANIUM FOR LITHIUM DRIFTING. Nucl. Instrum. Methods, 48
         (1967), 357-58.

     A description is given of a method of growing germanium cry-
stals which are effectively oxygen-free and suitable for the
fabrication of gamma ray spectrometers by the lithium drift process. Aluminum is used both as p-type dope and as a getter for
oxygen.

223. Eichinger, P. THEORETICAL LIMITS OF THE ENERGY RESOLUTION OF
         SOLID-STATE IONIZATION CHAMBERS, USING GE(LI) DETECTORS AS

AN EXAMPLE. (DIE THEORETISCHE GRENZE DER ENERGIE AUFLORE-
SUNG VON FESTK-OERPERIOMISATIONSKAMMERN AM BEISPIEL DES GE(LI)
DETEKTORS.) pp. 57-61 of "Soc. Europeenne de Protec. Contre
des Rayonnements Detectors in Radiation Protec. and Radiation
Measurement Tech. Sept. 1967." 1967. (In German)
                                          CONF-670545

Various models describing electron-hole pair formation in
semiconductor detectors by charged particles and gamma radiation
are discussed. Factors which limit the resolution of semiconductor detectors such as pulse rise-time and background and the
Fano factor are described. For a lithium-drifted germanium detector with a Fano factor <0.2 the attainable energy resolution
at half maximum intensity is 1.8 keV for 1 MeV particles ($\approx$0.2%)
and 5 keV for 10 MeV particles ($\approx$0.05%).

224. Elad, Lawrence and Nakamura, Michiyuki. HIGH-RESOLUTION BETA-
     AND GAMMA-RAY SPECTROMETER. IEEE Trans. Nucl. Sci., NS-14
     (Feb. 1967), 523-31.

     A discussion of a high-resolution semiconductor beta- and
gamma-ray spectrometer is presented. The spectrometer consists
of a silicon or germanium detector, low noise field effect transistor preamplifier, and a linear amplifier. The materials and
requirements for high-resolution performance are provided.

225. Elad, Lawrence and Nakamura, Michiyuki. HYPERCRYOGENIC DETECTOR-
     FET UNIT: CORE OF HIGH-RESOLUTION SPECTROMETER. IEEE Trans.
     Nucl. Sci., NS-15, no. 3 (June 1968), 477-85.
                                          Also report CONF-680207

     From the 11th Scintillation and Semiconductor Counter Sym-
posium, Washington, D. C.

     A hypercryogenic high-resolution x-ray and gamma-ray spec-
trometer consists of a lithium-drifted germanium detector and a
germanium-junction field effect transistor (JFET) preamplifier.
The detector is operated at its optimum temperature, which is
in the range of 10-30°K. The JFET's are operated at liquid helium temperature. This results in 0.28 keV fwhm preamplifier
resolution with a slope of 0.018 keV/pf. Resolutions of the
order of 0.4 keV are obtained with low capacitance detectors
for low energy x-rays. The gamma rays of [57]Co are measured with
0.68 keV fwhm. The described spectrometer was used to study
some aspects of the trapping effects in germanium detectors.

226. Edelstein, Norman; Hendry, David, and Michel, Maynard. ELEC-

TRONIC-INSTRUMENTATION AND SEMICONDUCTOR DEVICES. pp. 331-
42 of "Nuclear Chemistry Division Annual Report, 1968."
California University, Berkeley, Lawrence Radiation Lab.
Jan. 1969.   UCRL-18667

Topics include the development of a germanium counter system for charged-particle nuclear spectroscopy, germanium detector system development, computer control system for the field-
free spectrometer, control unit for linking pulse-height analyzers to a PDP-9 computer, and a special-purpose computer for linking pulse-height analyzers with a digital incremental plotter.

227. Ellis, R. LEAKAGE CURRENT AS A DIAGNOSTIC MEASUREMENT ON GER-
        MANIUM MATERIAL. pp. 81-99 of the "Proceedings of the Meet-
        ing on Special Techniques and Materials for Semiconductor
        Detectors, Ispra, Italy, 1968." June 1969.   EUR-4269

In the diagnostics of germanium material for use in lithium
drifted gamma spectrometers one of the useful parameters missing
from manufacturer's specifications is the room temperature diode
voltage current characteristic for a specified depleted volume.
Its implication on the processing of a typical lithium-drifted
gamma spectrometer is stated.

228. El-shishine, M. M. and Zobel, W. TEMPERATURE DEPENDENCE OF THE
        RESPONSE OF LITHIUM-DRIFTED GERMANIUM DETECTORS TO GAMMA
        RAYS. Oak Ridge National Lab., Tenn. Sept. 1966. 75p.
        (Thesis--Univ. Tenn., Knoxville.)
                                  ORNL-TM-1295; NASA-CR-83250

A report describing two Ge(Li) detectors, approximately
20 mm diameter and 3 mm thickness. The detectors were studied
at temperatures ranging from 85° to 160°K, using gamma ray sources
with energies varying from 279 to 1332 keV.

229. Emery, F. E. and Rabson, T. A. AVERAGE ENERGY EXPENDED PER
        IONIZED ELECTRON-HOLE PAIR IN SILICON AND GERMANIUM AS A
        FUNCTION OF TEMPERATURE. Phys. Rev., 140 (Dec. 13, 1965),
        A2089-93.

Measurements were made over a range of temperature of the a-
mount of ionization produced by the passage of ionizing radiation
through the junction of Ge and Si Li-drifted solid state detectors. These data are presented in the form of the average energy per hole-electron pair. A single model for the value of E,
the average energy per hole-electron pair, ions proposed by
Shockley in 1960. This model was extended to predict temperature

effects, upon E, Silicon and Ge(Li) drifted counters were designed for detecting 1 MeV electrons and for 5.5 MeV alphas o-
ver a temperature range of 4 to 300°K.

230. Emery, F. E. and Rabson, T. A.   TEMPERATURE DEPENDENCE OF AVER-
     AGE ENERGY PER PAIR IN SEMICONDUCTOR DETECTORS.   Trans. Nucl.
     Sci., NS-13, no. 1 (Feb. 1966), 48-52.

     A simple model for the value of ε, the average energy per elec-
     tron-hole pair, was proposed by Shockley in 1960.   This model was
     extended to predict temperature effects upon ε.   Si and Ge Li-
     drifted counters were designed and built for detecting 1 MeV
     electrons and/or 5.5 MeV alpha particles over a temperature
     range of 4°K to 300°K.   Techniques and equipment to measure the
     charged produced from a single incident ionizing particle with-
     in 0.4% were developed.   Data were taken for Si and Ge detectors
     at temperatures ranging from 20 to 300°K.   These data only qual-
     itatively agree with the simple model.   The evidence points to
     an additional effect over the ones involved by Shockley.   The
     room temperature values of E for the Si Li-drifted counters ob-
     served agree with those for Si surface-barrier detectors which
     were recently published.

231. Erramuspe, H. J. and Sinderman, J. E.   AUTOMATIC CONTROL OF THE
     DRIFTING PROCESS FOR SEMICONDUCTOR RADIATION DETECTORS.
     Rev. Sci. Instrum., 42, no. 2 (Mar. 1971), 373-76.

     An automatic Li drifting apparatus is presented which con-
     trols the temperature of the detector bath to maintain the leak-
     age current at a preselected value.   When the Li ions reach the
     back of the wafer the temperature is automatically reduced to
     room temperature, leaving or not, optionally, the bias voltage
     applied.   The need for constant supervision is therefore elimina-
     ted.

232. Erramuspe, H. J.   DEPLETION DEPTH DETERMINATION IN SEMICONDUCTOR
     RADIATION DETECTORS.   Nucl. Instrum. Methods, 78 (1970),
     175-6.

     A technique is described which allows a nondestructive deter-
     mination of the depletion depth while the crystal is under drift,
     with practically no interruption of the drifting process.

233. Euler, B. A. and Kaplan, S. N.   APPROXIMATE METHODS FOR CALCULA-
     TION OF INTRINSIC TOTAL-ABSORPTION AND DOUBLE-ESCAPE-PEAK
     EFFICIENCIES FOR GE(LI) DETECTORS.   IEEE Trans. Nucl. Sci.,

NS-17, no. 1 (Feb. 1970), 81-90. Also report CONF-691017-pt. 1

From 16th Nuclear Science Symposium, San Francisco, Calif.
Methods for calculation of intrinsic total-absorption and
double-escape-peak efficiencies of Ge(Li) detectors are described.

234. Ewan, G. T. and Tavendale, A. J. APPLICATION OF HIGH RESOLUTION
     LITHIUM-DRIFT GERMANIUM GAMMA-RAY SPECTROMETERS TO HIGH EN-
     ERY GAMMA RAYS. Nucl. Instrum. Methods, 26 (Feb. 1964),
     183-6.

     A description is given of the use of Li-drifted Ge detectors
for high resolution gamma spectroscopy. The detectors are used
in the following manner: the photon enters the detector and
annihilates into an electron-positron pair; the annihilation
quanta escape from the detector; and the total energy of the
electron-positron pair is lost in the detector. The detector
thus records a peak of $E-2m_0c^2$, where E is the photon energy and
m_0 is the electron rest mass. The method is useful only at high
photon energies. The detectors in this application have energy
resolutions of about 0.13%.

235. Ewan, G. T. EXPERIMENTS WITH GERMANIUM GAMMA SPECTROMETERS IN
     THE CHALK RIVER LABORATORY. Izv. Akad. Nauk. SSSR Ser. Fiz.,
     29 (July 1965), 1064-69. (In Russian)

     The properties of the Li-drifted p-i-n Ge detectors are
summarized including their design and layout, resolution for gamma rays of various energies, efficiency, resolving time, and
response to high-energy gammas. The performance of these detectors in recording the gamma decay scheme of ^{134}Cs in measuring
internal conversion coefficients, in determining high-levels in
deformed enen-enen nuclei, in measuring the lifetime of nuclear
states by the recoil-nucleus method is described.

236. Ewan, G. T.; Andersson, G. I.; Bartholemew, G. A., and Lither-
     land, A. E. GAMMA-RAY POLARIZATION MEASUREMENTS WITH A SIN-
     GLE GE(LI) DETECTOR. Phys. Lett., 29B (June 9, 1969), 352-4.

     A single rectangular Ge(Li) detector, $4 \times 2.5 \times 0.35$ cm^3 was
measured. It was used to study the polarization of the prominent gamma rays emitted in the ^{100}Mo$(\alpha,2n)^{102}$Ru reaction.

237. Ewan, G. T. and Tavendale, A. J. HIGH-RESOLUTION STUDIES OF
     GAMMA RAY SPECTRA USING LITHIUM-DRIFT GERMANIUM GAMMA RAY

SPECTROMETERS. Can. J. Physics, 42 (Nov. 1964), 2286-2331.
Also report AECL-2079

This paper discusses the use of Ge(Li) drift p-i-n diodes
as high-resolution gamma ray spectrometers.

238. Ewan, G. T.; Malm, H. L., and Fowler, I. L. LITHIUM-DRIFTED
GERMANIUM GAMMA-RAY SPECTROMETERS AT CHALK RIVER. SOME
RECENT WORK. pp. 102-132 of "Panel on Lithium-Drifted
Germanium Detectors Proceedings...." Vienna, International
Atomic Energy Agency, 1966.

A review of work with Ge(Li) detectors at Chalk River.

239. Ewan, G. T.; Malm, H. L., and Fowler, I. L. RECENT WORK WITH
GE(LI) GAMMA-RAY SPECTROMETERS AT CHALK RIVER. Presented
at the IAEA Panel Meeting on the use of Germanium-Lithium-
Drift gamma-ray Detectors for Res. in Nuclear Physics, Vienna,
1966. June 1966. 58p.   Also report AECL-2607

A review is given of some recent work with Ge(Li) detectors
at Chalk River. The development and properties at large volume
coaxial detectors are discussed briefly and typical results of
measurements of the resolution, efficiency, and peak-to-total
ratios for these detectors are presented.

240. Ewan, G. T.; Graham, R. L., and MacKenzie, I. K. USE OF GE(LI)
DETECTORS IN GAMMA-GAMMA COINCIDENCE EXPERIMENTS. IEEE Trans.
Nucl. Sci., NS-13, no. 3 (June 1966), 297-305.
Also report AECL-2565

Large volume Ge(Li) detectors have been used in a gamma-gamma
coincidence arrangement employing two Ge(Li) detectors. The first
part of this part reports the results of studies of the timing
properties of planar and coaxial diodes and the time resolution
curves that can be obtained. For two cylindrical coaxial diodes
the full-width at half maximum of the time resolution curve was
14 nsec for 511 keV.

241. Ewan, G. T. THE USE OF GERMANIUM LITHIUM-DRIFTED P-I-N DIODES
AS HIGH RESOLUTION GAMMA-RAY SPECTROMETERS. Chalk River,
Ont., Canada. April 1964. 24p.   AWCL-1960

Text of Invited Talk at the Annual Meeting of American Physical Society, New York, Jan. 28, 1964.

Lithium-drift germanium diodes have been used to study gamma-
ray spectra from the x-ray region up to 9 MeV. The resolution
(FWHM) obtained at 122 keV was 3.1 keV and 7285 keV was 9.8 keV.
Intrinsic efficiency curves are given for an 18 mm diameter detector with a 3.5 mm depletion depth. A brief description is given
of a 3-crystal pair spectrometer using a lithium-drift germanium
diode as the central crystal.

242. EXPERIMENTAL FACILITIES, TECHNIQUES, AND INSTRUMENTATION. pp.
      181-316 of "Nuclear Technology Branch Yearly Progress Re-
      port for Period Ending June 30, 1968." Dec. 1968.
                                                        IN-1218

243. Fabjan, C. W.; Kenawy, M., and Rauch, H. STUDY OF SEMICONDUCTOR
      SURFACE BARRIER DETECTORS AT 4.2 DEG K. Z. Phys., 188 (1965),
      378-84. (In German)

The behavior of semiconductor surface barrier detectors was
studied in the temperature range from 300 to 4.2°K. Measurements of the resolution for gamma and β particles are given.
The mean energy E required for producing an electron-hole-pair
in Si is found to be 3.54±0.10 ev at 279°K, and 3.72±0.11 ev at
78°K and 4.2°K respectively. Measured rise time of detector
impulses is compared with theoretical values.

244. Fairstein, E. CHARACTERISTICS AND REQUIREMENTS OF PULSE AMPLI-
      FIERS FOR USE WITH GERMANIUM (LITHIUM) DETECTORS. pp. 411-
      20 of "Semiconductor Nuclear Particle Detectors and Circuits."
      Brown, W. L., ed. Washington, D. C., National Academy of
      Sciences, 1969.   Also report CONF-670520

From Conference on Semiconductor Nuclear Particle Detectors
and Circuits, Gatlinburg, Tenn.

Describes recent progress in the development of amplifiers
for high-resolution Ge(Li) detectors.

245. Fanger, U. TWO-PARAMETER GAMMA-GAMMA COINCIDENCE ANALYSIS WITH
      A GE(LI)-NAI(TL) SPECTROMETER AT THE KARSRUHE FR-2 REACTOR.
      Kernforschungszentrum, Karlsruhe (West German) Instut fuer
      Angewandte Kernphysik. Jan. 1969. 13p. (In German)
                                                        KFK-887

The coincidence combination of Ge(Li) diodes with NaI(Tl)
detectors represents an efficient spectrometer especially in
the gamma energy range from 100 keV to 3 MeV. The coupling of

the apparatus to the Karlsruhe computer system MIDAS makes possible two-parameter measurements with full channel resolution. The advantages resulting lie in the simultaneous detection of several coincidences and in the optimum positioning of the coincidence window in spectra with good counting statistics at the end of a measurement.

246. Festag, J. G. SOME GAMMA SPECTRA OF ACTIVATED CYCLOTRON PARTS. (EINIGE GAMMA SPEKTREN AKTIVIERTER ZYKLOTRONTEILE). pp. 156-163 of "Soc. Europeenne de Protec. Contre des Rayonnements Detectors in Radiation Protec. and Radiation Measurement Tech. Sept. 1967." 1967. (In German)

The use of a Ge(Li) detector for the detection and determination of the radio-induced activity of cyclotron parts is described. The results obtained were compared with those obtained using a NaI(Tl) crystal. The comparison showed that with a Ge(Li) detector the long-lived activities important for radiation protection problems can be detected or separated much earlier than with NaI(Tl) crystals.

247. Fettweis, P. THE FABRICATION OF GE(LI) DETECTORS. Centre d' Etude de l'Energie Nucleaire, Brussels, Belgium. Sept. 1970. 27p.   BLG-447

The fabrication of lithium-drifted germanium detectors for use in gamma-ray spectroscopy is described. The performances of the different types such as planar and one and two open-ended coaxial detectors are briefly discussed in view of their applications in straight spectroscopy and timing circuits.

248. Fettweis, P. and Vervier, J. GAMMA-RAY SPECTROSCOPY OF SHORT HALF-LIFE ISOTOPES WITH A LITHIUM-DRIFTED GERMANIUM DETECTOR. Censck. Mol. Belg. 1966. 10p.
                                    Also report AED-CONF-65-170-46

From American Physical Society, Summer Meeting, New York.

The decay spectra of tantalum-182, isomeric indium-116, and vanadium-52 are found using a lithium-drifted germanium detector. The performance of this detector is described. Also the decay spectra of the three isotopes studied and the energy levels of isomeric tantalum-182, tin-116, and chromium-52 are given.

249. Fettweis, P. SOME FUNDAMENTAL REMARKS ON GE(LI) DETECTORS. Paper 1 of "The proceedings of the Meeting on the Applica-

tions of Ge(Li) Detectors in Science, Technology, Medicine
and Industry, Brussels. Oct. 20, 1967." 1967. 15p. BLG-425

A description of the operation of lithium-drifted germanium
detectors for gamma spectroscopy is presented. The fundamental
properties and operation of the lithium—drifted germanium detectors are compared to those of an ionization chamber.

250. Fiedler, H. J.; Hughes, L. B.; Wall, B. J.; Kenneth, T. J., and
     Prestwich, W. V. LARGE VOLUME LITHIUM-DRIFTED GERMANIUM
     GAMMA-RAY DETECTORS. Nucl. Instrum. Methods, 40 (1966),
     229-234.

A new electroplating method was used for the fabrication of
large-volume Ge(Li) detectors in "wrap around" configuration.
The properties and performance of 6 cm^3 and a 10 cm^3 detector
were investigated. Some applications are described.

251. Filby, Royston H. and Haller, William A. ACTIVATION ANALYSIS OF
     GEOCHEMICAL MATERIALS USING GE(LI) DETECTORS. pp. 339-346
     of "Modern Trends in Activation Analysis." Vol. 1. Washing-
     ton, D. C., National Bureau of Standards, 1969.
                                             CONF-681003-v.1

The work described in this paper was undertaken to extend
the range of elements which may be determined in rocks, minerals,
and meteorites. To include nuclides with half lives in the
range 2 minutes to 40 hours, the method developed consists of
three parts. Short neutron irradiation of samples and standards
followed by rapid transfer to the counting system permits the
determination of Al, Mn, Na, V, Ti, Ca and in certain cases, Mg.
in rocks and meteorites. Long irradiation followed by a one
month decay period allows the determination of Sb, Co, Zn, Sc,
Cr, Eu, Th, U, Ba, Ta, and Fe. The high Na-24 activity in most
irradiated silicate rocks and meteorites masks several activation products of interest. A method has been developed for the
determination of these nuclides by the dissolution of the irradiated sample followed by removal of Na-24 on a sodium specific
inorganic ion-exchanger. The separation of Na-24 is simple and
negligible absorption of other elements of interest has been
observed.

252. Fischer-Colbrie, Erwin; Brown, T. G., and Friensehner, A. V.
     THE EFFECTS OF HEAT TREATMENTS OF GERMANIUM ON RESISTIVITY,
     PHOTOCONDUCTIVE DECAY AND ON LITHIUM DRIFT PROPERTIES.
     1968. 13p.   UCRL-71082-Rev. 1

Presented at 15th Nucl. Sci. Sym., Montreal, 1968.

It was found that in contrast to literature, thermally produced acceptors cannot be reconverted if vacuum used instead of
a gas during anneal treatments. Vacuum heat treatments also lead
to a stronger decrease of the photoconductive decay time than a
gas atmosphere does. Close to theoretical lithium drift rates
considerably decrease in heat treatments, poor drift rates
can be improved. The effects of quenching differ distinctly
from those of slow cooling. The surface exposed to vacuum can
be assumed to be a more abundant source of single vacancies due
to a lower density of sorption engagements.

253. Fleck, C. M. and Niederstaetter, W. A CRYOSTAT FOR GE(LI)-DE-
TECTORS WITH INTERNAL LIQUID CIRCULATION AND A SWIVELLING
COLD FINGER. Nucl. Instrum. Methods, 66 (Dec. 15, 1968),
304-8.

The description and operations of a cold finger cryostat
are presented.

254. Forest, H.; Hugeut, M., and Ythier, C. PULSE SHAPE DISCRIMINA-
TION IN COINCIDENCE SPECTROMETERS USING A GE DETECTOR. Nucl.
Instrum. Methods, 74, no. 2 (Oct. 1969), 325-8. (In French)

A simple and accurate pulse shape discrimination method is
described for coincidence spectrometers using Ge(Li) detectors.
Gamma ray pulses degraded by electron interaction in Ge(Li) detector's dead layers are rejected within 40% by delaying the
Ge(Li) crossover signals. Spectra given by a Ge(Li)-NaI(Tl)
three crystal pair spectrometer with and without rejection are
shown. Results are compared with elaborated and much more ex-
pensive methods.

255. Forsyth, R. S. and Blackadder, W. H. A GAMMA SCANNER USING A
GE(LI) SEMICONDUCTOR DETECTOR WITH THE POSSIBILITY OF OPER-
ATION IN THE ANTICOINCIDENCE MODE. Aktiebolaget Atomenergi,
Studsvik, Sweden. Apr. 1970. 21p.   AE-393

A fuel element transport flask has been modified as a facility for gamma scanning irradiated fuel elements up to a length
of 75 cm. By means of a Ge(Li) semi-conductor detector, satis-
factory activity profiles along the specimens have been obtained,
permitting the location of individual fuel pellets. An annular
plastic detector surrounding coincidence mode, and reduction of
the Compton continuum by about 50% has been obtained.

256. Fouan, J. P. and Passerieux, J. P. A TIME COMPENSATION METHOD
        FOR COINCIDENCE USING LARGE COAXIAL GE(LI) DETECTORS. Nucl.
        Instrum. Methods, 62 (1968), 327-29.

        A method is described to compensate for the time jitter due
     to slow and variable pulse rise time. This allows to work with
     any type of detector and gives a time resolution $2\gamma < 10$ns, for
     all gamma energies down to 100 keV.

257. Fowler, I. L. and Toone, R. J. PRELIMINARY TESTS ON ENCAPSU-
        LATED LITHIUM-DRIFT GERMANIUM GAMMA-RAY SPECTROMETERS MADE
        BY RCA VICTOR RESEARCH LABORATORIES, MONTREAL. July 1964.
        14p.   AECL-2569; GPI-59

        Tests were carried out on three encapsulated germanium lith-
     ium-drift gamma ray spectrometers. The detectors are approxi-
     mately 2 cm^2 in area and 2 mm depletion (sensitive) depth and
     are encapsulated in modified standard steel transistor cans.
     After a straight-forward electrical clean-up procedure the de-
     tectors were cooled to liquid nitrogen temperatures (77°) and
     gamma spectra taken. All three detectors gave resolutions of
     2.6 to 2.7 keV (fwhm) for 122 keV gamma rays and 6.3 keV for
     1173 keV gamma rays. The detectors were found to operate satis-
     factorily after storage for several days in a freezer at -22°C
     without further clean-up. One detector was operated, cooled to
     near 77°K with liquid nitrogen, in an insulated mount with no
     vacuum system. Such an arrangement is only possible with encap-
     sulated detectors of this type and should prove convenient for
     many users for short term experiments. Details of this non-
     vacuum are given.

258. Fox, R. J. DETERMINATION OF OXYGEN IN GERMANIUM BY LITHIUM
        PRECIPITATION. pp. 198-200 of "Semiconductor Nuclear-Par-
        ticle Detectors and Circuits, Proceedings." Brown, W. L.,
        ed. Washington, D. C., National Academy of Sciences, 1969.

259. Fox, R. J.; Williams, I. R., and Toth, K. S. A GE(LI) GAMMA
        RAY SPECTROMETER. Nucl. Instrum. Methods, 35 (1965), 331-3.

        A gamma- and x-ray spectrometer is described in which a
     lithium drifted germanium crystal immersed in liquid nitrogen
     is used. Photopeak detection efficiency curves obtained for
     Ge(Li) and NaI(Tl) crystals using x and gamma sources are in-
     cluded, together with ^{207}Bi gamma calibration spectra obtained
     with both types of crystals. Photon spectra obtained from gad-
     olinium isotopes using Ge(Li) detectors demonstrate the use of
     the detector both in the conventional manner and as a sum spec-
     trometer.

260. Fox, R. J. LITHIUM DRIFT RATES AND OXYGEN CONTAMINATION IN
     GERMANIUM. IEEE Trans. Nucl. Sci., NS-13, no. 3 (June 1966),
     367-69.

     The unpredictable variations in lithium drift rates of vari-
     ous commercially available germanium crystals have made the pro-
     duction of germanium gamma-ray detectors rather difficult. Some
     as-received germanium ingots have shown diffusion constants D re-
     duced by nearly three decades. Because drift time varies inverse-
     ly with D, the time required to drift such germanium to make
     large-volume detectors can become too long to be practical. The
     drift rates and lithium precipitation kinetics of such germanium
     were investigated to determine if they were correlated.

261. Fox, R. J. METHOD FOR PREPARATION AND ENCAPSULATION OF GERMANIUM
     GAMMA RAY DETECTORS. July 2, 1968.   U. S. Pat. 3,390,449

262. Fox, R. J. PRODUCTION AND ENCAPSULATION OF LITHIUM GERMANIUM DE-
     TECTORS. pp. 44-45 of "Oak Ridge National Laboratory, In-
     strumentation and Controls Division, Annual Progress Report,
     Sept. 1, 1965." 1965.   ORNL-3875

263. Franke, H. F. FLEXIBLE DETECTOR CRYOSTAT. Nucl. Instrum. Meth-
     ods, 72 (1969), 107-8.

     A cryostat for nonencapsulated Ge(Li) semiconductor radiation
     detectors is described. The connection between cryostat and li-
     quid nitrogen reservoir is flexible, so that the cryostat can be
     used in all positions between verticle and horizontal (right
     angle).

264. Freck, D. V. and Wakefield, J. GAMMA-RAY SPECTRUM OBTAINED WITH
     A LITHIUM DRIFTED P-I-N JUNCTION IN GERMANIUM. Nature, 193
     (1962), 669.

265. Freeman, J. M. and Jenkin, J. G. THE ACCURATE MEASUREMENTS OF
     THE RELATIVE EFFICIENCY OF GE(LI) GAMMA-RAY DETECTORS IN THE
     ENERGY RANGE OF 500 TO 1500 KEV. 1965. 26p.   AERE-R-5142

266. French, W. R.; LaShure, R. E., and Curran, J. LITHIUM DRIFTED
     GERMANIUM DETECTORS. Amer. J. Physics, 37, no. 1 (1969),
     11-22.

     Lithium-drifted germanium detectors may be profitably used

in variety of experiments in undergraduate physics laboratories.
We have attempted to assemble from literature and our own experience an introduction to these devices. Included are some basic
experiments relating to the properties of the detectors which are
interesting in themselves, but also practical in terms of routine
detector and electronic system evaluation.

267. Freund, Hans-Ulrich. STUDY OF GAMMA RAYS OF AG 110 AFTER NEUTRON
     CAPTURE. (UNTERSUCHUNG DER GAMMASTRAHLUNG VON AG 110 NACH
     NEUTRONENEINFANG.) Technishce Hochschule Munchen (W. Ger.)
     Fakultaet fuer Allgemeine Wissenschaften. 1967. 53p.
     Thesis. (In German)

     Germanium-lithium detectors were used to measure a gamma ra-
     diation from captured neutrons in silver 109. Transfer electron
     intensities were analyzed to determine spin and parity of the
     various energy levels during isotopic decay.

268. Friant, Alain. NONDISPERSIVE X-RAY SPECTROMETRY WITH SI(LI) AND
     GE(LI) DETECTORS. Commissariat a l'Energie Atomique, Saclay,
     France. Centre d'Etudes Nucleaires. June 10, 1970. 41p.
     (In French)   CONF-700619-1

269. Friedland, S. S. GAMMA ASSAY WITH LITHIUM DRIFTED GERMANIUM SEMI-
     CONDUCTOR NUCLEAR DETECTORS. Vienna, International Atomic
     Energy Agency, 1965. 12p.   CONF-650507-33

     From IAEA Symposium on Radioisotope Sample Measurement Tech-
     niques in Medicine and Biology, Vienna.

     The characteristics and applications of the semiconductor nu-
     clear radiation detector which has been fabricated from germanium
     and compensated to high sensitive volume for gamma ray detection
     are presented. Detector thicknesses of 1 to 4 mm are readily
     fabricated whereas 6 mm and greater are low yield devices. Gamma
     resolutions of less than 10 keV are obtainable under standard
     laboratory conditions and less than 3 keV under ideal conditions.
     Such high resolutions will have immediate effect on gamma assay
     as it may prove possible in many applications to remove the need
     for chemical separation. Results with ^{207}Bi, ^{166}Ho, and other
     sources will be shown.

270. Friedland, S. S. SOLID STATE RADIATION DETECTOR WITH SEPARATE
     OHMIC CONTACTS TO REDUCE LEAKAGE CURRENT. Jan. 7, 1964.
                                                U.S. Pat.3,117,229

271. Fry, Edward S.; Palms, John M., and Day, Robert B. CALCULATION
     OF THE PULSE HEIGHT RESPONSE OF GE(LI) SEMICONDUCTOR COUNTERS.
     Los Alamos Scientific Lab., N. Mex. 1966. 33p.   LA-3456

     A Monte Carlo calculation was used to determine the pulse-
     height response of Ge(Li) detectors to gamma rays. A set of runs
     for three different crystal sizes and six different energies
     was made. The crystals were cylinders of length A and diameter A,
     where A was successively given the values of 1 cm, 1 in., and
     3 in. The source was a point source 10 cm from the end of the
     crystal, and the gamma-ray energies in MeV were 0.279, 0.412,
     1.114, 2.7535, nnd 4.432.

272. Fubini, A.; Terrasi, F., and Vata, I. A METHOD TO IMPROVE THE
     ENERGY CALIBRATIONS OF A GE(LI) SPECTROMETER. Nucl. Instrum.
     Methods, 92, no. 2 (March 15, 1971), 309-11.

     Discusses the method of improving precision energy calibra-
     tion of a Ge(Li) detector.

273. Fuelle, R.; Netzband, D., and Schlott, K. PARTICLE DISCRIMINA-
     TION WITH SEMICONDUCTOR DETECTORS. pp. 84-97 of "Internation-
     ale Arbeitstagung Herstellung und Anwendung Von Halbeiterde-
     tektoren, 23-28 Sept. 1963 Rossendorf." (International Con-
     ference on the Preparation and Utilization of Semiconductor
     Detectors, Sept. 23-28, 1963, Rossendorf.) Oct. 1963. (In
     German)   Also report ZFK-PHA-12

     The principle of the method of thin detectors was explained.
     Special semiconductor detectors are appropriate for this method.
     The applicability of the method in nuclear reactions was dis-
     cussed in general. It appeared that the range of applicability
     was significantly enlarged when a combination of dE/dx counters
     and thin semiconductor detectors was used. The electronics were
     described and practical experiences and measurements results in
     the reaction $^{10}B(d, t_0)^9B$ were reported.

274. Fuller, C. S. and Severiens, J. C. MOBILITY OF IMPURITY IONS IN
     GERMANIUM AND SILICON. Phys. Rev., 96, no. 1 (Oct. 1, 1954),
     21-24.

     Lithium has been shown to migrate as a singly-charged posi-
     tive ion in single crystals of both Ge and Si in temperature
     ranges of 15-600°C, respectively. The mobility of the Li+ in
     crystalline Ge and Si has been measured as a function of tempera-
     ture. Through the use of the Einstein relation between diffusion

constant and mobility, values of the diffusion constants in
cm^2/sec of Li in Ge and Si are obtained as follows: D=25x10-
4exp(-11800)/RT for $D = 23x10-4exp(-15200)/RT$ for Si, in satis-
factory agreement with previously published results on the thermal
diffusion of Li.

275. FURTHER DEVELOPMENT OF SEMICONDUCTOR DETECTORS FOR GAMMA TRAN-
        SIENT DETECTION. FINAL REPORT. Solid State Radiation, Inc.,
        Los Angeles, Calif. 1966. 61p.   UCRL-13272

        The fabrication of semiconductor detectors for gamma tran-
    sient detection is discussed. Germanium, gold-doped silicon,
    gallium arsenide as well as silicon diodes were fabricated and
    evaluated. Methods of optical dry running of semiconductor were
    investigated and a pulse xenon flash lamp was designed for filed
    use. Finally, protolype silicon diodes suitable for application
    in a high-vacuum photodiode were designed and fabricated.

276. Gallardo Villegas, Raul. DEVELOPMENT OF LITHIUM-ACTIVATED GER-
        MANIUM SEMICONDUCTOR NUCLEAR DETECTORS. Universidad Nacion-
        al Autonoma de Mexico, Mexico City. 1970. 72p. (In Span-
        ish)   Thesis   Also report NP-18180

        The construction of coplanar lithium-drifted germanium de-
    tectors is presented. The interaction of nuclear radiation with
    matter, which is necessary in order to understand the function-
    ing of semiconductors, is discussed. The properties and the
    theory of semiconductors necessary for detectors and the proper-
    ties of the diffusion mechanism used to obtain ions in a n-p
    junction are considered. The operation and some uses of vari-
    ous types of detectors are treated in general form and the ad-
    vantages and disadvantages of Li-drifted Ge detectors are given.
    The technique used in the construction of Ge(Li) detectors is
    presented, and the results obtained in the manufacture of these
    detectors is given.

277. Gallmann, A. and Molinari, M. A. CONSTRUCTION OF -I-COMPENSATED
        GE DETECTORS. RC Accad. Naz. Lincei, (Italy), 45, no. 5
        (Nov. 1968). (In French)

        Several Ge(Li) detectors were studied by using the gamma-ray
    scanning technique. A particular example is that of a single-
    open end detector, the technique, the performances of which have
    been substantially increased after cutting the bad region. In
    order to improve the photopeaks-to-Compton ratio of coaxial detec-
    tors, a spark machine was used for removing the central P-type
    core. Two such cases are presented.

278. Gallmann, A. and Molinari, M. A. DEVELOPMENT OF A DETECTOR OF
     GERMANIUM COMPENSATED WITH LITHIUM. Rev. Phys. Appl., 4
     (June 1969), 285-6. (In French)
                                   Also report CONF-681208-Suppl.

     From Conference on Experimental Methods in Nuclear and Par-
     ticle Physics, Strasbourg, France.

     Several Ge(Li) detectors were studied by using the gamma-ray
     scanning technique. A particular example is that of a single-
     open end detector, the performances of which have been substan-
     tially increased after cutting the bad region. In order to im-
     prove the photopeaks-to-Compton ratio of coaxial detectors, a
     spark machine was used for removing the central P-type core.
     Two such cases are presented.

279. GAMMA SPECTROSCOPY WITH SEMICONDUCTORS. Naturwiss. Rundsch., 21
     (Feb. 1968), 71-2. (In German)

     The characteristics of the semiconductor spectrometer are
     described. This spectrometer is based on the fact that ioniza-
     tion in the crystal produces an electric pulse whose strength
     is proportional to the energy of the radiation. The electron
     yield is considerably greater than in a gas ionization chamber.
     In principle, it is similar to a photoelectric meter. Semicon-
     ductors such as Si and Ge can be used. In Ge a 300 keV photo-
     electron produces 10^5 ion pairs, which is several times those
     produced in other devices. Because of this it is possible to
     measure gamma energies of 1 keV accurately, while the breadth of
     the spectral peak from a proton multiplier is ~25 keV. There is
     some problem in producing physically pure Ge with a perfect cry-
     stal structure so that interference voltages will be excluded.
     The largest component part is the container for liquid N_2 used
     for cooling. The semiconductor spectrometer is readily used with
     a digital computer.

280. Gammel, Reinhold. MEASURING AND ANALYZING METHODS WITH MULTI-
     CHANNEL ANALYZER VAS 4096. Siemens Rev., 37 (Dec. 1970),
     628-31.

281. Ganner, P. and Rauch, H. PERFORMANCE OF A GE(LI)-DETECTOR IN HIGH
     MAGNETIC FIELDS. Nucl. Instrum. Methods, 76 (1969), 295-300.

     An encapsulated Ge(Li)-gamma-detector (2 cm^3 of active volume)
     has been placed into a strong magnetic field of a superconducting
     magnet (up to 60 KOe) and its effect on energy resolution and full
     energy peak efficiency was measured in the energy range from 40 to

1333 keV at 6 different detector bias voltages. The fwhm was found to be increased by increasing magnetic field and decreasing bias voltage. The full energy peak efficiency was sharply decreasing in the field. A cryostat system was constructed for operating the Ge(Li)-gamma-detector in the field of the superconducting magnet. A brief phenomenological explanation of the measured results is given.

282. Gehrke, R. J.; Cline, J. E., and Heath, R. L. DETERMINATION OF RELATIVE PHOTOPEAK EFFICIENCY AND SYSTEM LINEARITY FOR GE(LI) GAMMA-RAY SPECTROMETERS. Nucl. Instrum. Methods, 91 (1971), 349-56.

A simple precise method is described for determining both the relative detection efficiency and the linearity of a Ge(Li) spectrometer. The method employs three readily available radioactive sources--^{75}Se(120d), ^{82}Br(35h), and ^{56}Co(77d)--that emit gamma rays covering the energy range 90-3500 keV. The gamma-ray energies and relative intensities for these isotopes have been precisely measured. A method is suggested to reproducibly set the "effective zero" of the analyzer using the above gamma-ray sources.

283. GEOMETRY EFFECTS IN A GAMMA JUNCTION RADIATION DETECTOR. Note no. 90. Ministere des Armees, Arcueil, (France). Laboratoire Central de l'Armement. Nov. 1968. 34p. (In French)
NP-18684

An attempt was made to determine the optimum dimensions compatible with the industrial manufacturing of Ge junction detectors of parallelepiped form for detecting 15 to 150 keV x-rays. The energy loss by the radiation in the bulk of Ge was evaluated. The process used for the numerical calculation was a simulation accurately reproducing the physical phenomena occurring during the detection and based on a Monte-Carlo method. The advantages of voluminous detectors in detecting the maximum number of photons are indicated.

284. Giauque, Robert D. RADIOISOTOPE SOURCE: TARGET ASSEMBLY FOR X-RAY SPECTROMETRY. Anal. Chem., 49 (Nov. 1968), 2075-7.

A radioisotope source-target assembly for x-ray fluorescence analysis using lithium-drifted germanium and silicon as detectors is described. Some of the radioisotopes used as sources are ^{241}Am, ^{109}Cd, ^{125}I, and ^{57}Co.

285. Gibbons, P. E. CHARGE COLLECTION IN GERMANIUM DETECTORS. pp.

31-36 of the "Proceedings of the Meeting on Special Techniques and Materials for Semiconductor Detectors, Ispra, Italy, 1968." June 1969.   EUR-4269

286. Gibbons, P. E. THE EFFECT OF CLUSTERS OF RECOMBINATION CENTRES ON LIFETIME MEASUREMENT. pp. 25-30 of the "Proceedings of the Meeting on Special Techniques and Materials for Semiconductor Detectors, Ispra, Italy, 1968." June 1969.
                                                    EUR-4269

287. Gibbons, P. E. THE DEVELOPMENT AND APPLICATION OF LI-DRIFT GERMANIUM DIODES AT A.E.R.E. Atomic Energy Research Establishment, Harwell, (England). May 1966. 24p.   AERE-R-5206

A state-of-the-art account of detector fabrication and developments in cryostat design and low noise head amplifiers, using field effect transistors is presented. Present applications and future requirements are summarized.

288. Gibbons, P. E.; Howes, J. H., and Pyrah, S. AN ENCAPSULATED LITHIUM DRIFTED GERMANIUM DIODE FOR GAMMA RAY SPECTROMETRY. Nucl. Instrum. Methods, 45 (1966), 322-24.

289. Gibbons, P. E. and Howes, J. H. GAMMA RAY SPECTROMETER SYSTEMS USING LITHIUM DRIFTED GERMANIUM DETECTORS. Atomic Energy Research Establishment, Harwell, England. Electronics and Applied Physics Div. Mar. 1968. 37p.   AERE-R-5703

The basic features of these detectors and their operational requirements are described. Operation of these detectors gives rise to some problems which must be solved if their potentials are to be fully and reliably achieved. This has involved the design of various types of cryostats, and the development of low-noise pulse amplifiers which can be matched to a detector to give an optimum response. This work leads to the concept of a spectrometer system, constructed from a basic set of components, designed to meet specific counting requirement. Several spectrometer systems were built and are described in detail. The degree of resolution obtainable has opened up many new areas of work in gamma spectrometry.

290. Gibbons, P. E.; Howes, J. H., and Owen, R. B. IMPROVEMENTS IN THE PERFORMANCE OF GERMANIUM DETECTORS FOR GAMMA-RAY SPECTROMETRY. pp. 152-8 of "Nucleonic Instrumentation." London, Institution of Electrical Engineers, 1968.   CONF-680939

From Conference of Nucleonic Instrumentation, Reading, England.

Over the last five years gamma ray spectrometry has been revolutionized by the development of lithium-drifted germanium detectors. These detectors have found application in many areas where precise energy measurement or discrimination between gamma rays of similar energies is important. The successful application of the detectors requires that careful attention be paid not only to the detector itself, but also the low temperature cryostat design, the low noise head amplifier, and the ability of the pulse analysing equipment to perform at high count rates with low distortion of the spectrum. The development programme at A.E.R.E. has aimed at optimizing all these features. In this paper, further developments which are relevant to a more complex exploitation of germanium detectors are discussed. The topics cover cryostat design, low noise amplifier arrangements, and some relevant detector results.

291. Gibbons, P. E. and Iredale, P. ON THE ACCURACY OF ACCEPTOR COMPENSATION BY LITHIUM ION DRIFT. Nucl. Instrum. Methods, 53 (1967), 1-6.

292. Gibbons, P. E. and Owen, R. SEMICONDUCTOR DETECTORS FOR GAMMA-RAY SPECTROSCOPY. Atomic Energy Research Establishment, Harwell, England. July 1965. 14p.   AERE-M-1602

   The use of semiconductor detectors in the field of charged particle spectroscopy is discussed. The performance of silicon and germanium semiconductors for gamma-ray spectroscopy is discussed. The construction of the germanium detectors and prospects for high atomic number semiconductors are included.

293. Gibbons, P. E. and Howes J. H. SMALL GERMANIUM DETECTOR WITH NEAR PERFECT COLLECTION. Nucl. Instrum. Methods, 73 (1969), 221-2.

   This article describes the performance of a small Ge(Li) detector.

294. Gibson, J. A.; Gibson, B., and Burt, A. K. CALIBRATION OF A PORTABLE GAMMA-RAY SPECTROMETER. Health Physics, 15, no. 4 (Oct. 1968), 333-4.

295. Giesler, G. C.; McHarris, Wm. C.; Warner, R. A., and Kelly, W. H. SPURIOUS PEAKS PRODUCED BY COMPTON SCATTERING IN GE(LI)-GE(LI)

COINCIDENCE GAMMA-RAY SPECTROMETER SYSTEMS. <u>Nucl. Instrum.</u> <u>Methods</u>, 91 (1971), 313-20.

Compton scattering between Ge(Li) γ-ray detectors in coincidence experiments was investigated. Such scattering can generate false peaks that can be mistaken for photopeaks. The effects varying gate position, gate width, background subtraction, and angle between detectors are discussed, and suggestions are made for recognizing and minimizing spurious effects.

296. Girard, J.; Samama, R.; Carlos, P.; Maier, B. P., and Perrin, G. PAIR AND ANTI-COMPTON SPECTROMETER. <u>Rev. Phys. Appl.</u>, 4 (June 1969), 261-2. (In French)   Also report CONF-681208

From Conference on Experimental Methods in Nuclear and Particle Physics, Strasbourg, France.
A gamma ray spectrometer working either in the anti-Compton or in the pair mode is described. It is composed by a 10 cm^3 Ge(Li) detector surrounded by a large NaI(Tl) annulus which is divided in 6 sectors. The pulses coming from these sectors are combined and used to condition the analysis of the pulses detected by the Ge(Li) detector.

297. Girardi, Francesco; Giampoalo, Guzzi, and Pardy, Jules. APPLICATION OF LITHIUM-DRIFTED GERMANIUM SOLID STATE DETECTORS TO THE DETERMINATION OF HAFNIUM IN ZIRCONIUM OXIDE BY ACTIVATION ANALYSIS. <u>Radiochim. Acta</u>, 4 (June 1965), 109-10.

Samples (about 0.1 g of zirconium oxide) were irradiated for two hours in the reactor in a flux of about 2×10^{13}/sec. The irradiated capsule was then left for 2 to 7 days, to allow for the decay of the short-lived interfering radioactivies especially that of ^{97}Zr and ^{97}Nb. Then the radioactivity was measured by means of a lithium-drifted germanium detector. The gamma spectrum of an irradiated sample of zirconium oxide containing 90± ppm of hafnium is shown.

298. Gizon, J. SOLID-STATE DETECTOR SPECTROMETERS FOR DETERMINATION OF MULTIPLE ORDERS OF NUCLEAR TRANSITIONS. <u>Nucl. Instrum.</u> <u>Methods</u>, 74 (1969), 213-18. (In French)

A description of an apparatus for the measurement of internal conversion coefficients is presented. The determination of multipole orders of nuclear transitions by simultaneous countings of electron spectrum with silicon detector and gamma spectrum with germanium detector is examined. Application to the decays of ^{171}Hf and ^{171}Lu is included.

299. Glasow, P. GE(LI) WELL-TYPE DETECTORS FOR MEASURING LOW GAMMA
     ACTIVITIES. Nucl. Instrum. Methods, 80 (1970), 141-5.
     (In German)

     When making measurements with Ge(Li) well-type detectors it
was found that contrary to expectations, the pulse rate increased
by more than the factor of 2 when changing from a 2π geometry to
a nearly 4π geometry in the well. Gamma rays backscattered during
Compton processes are also detected and increase the efficiency.
This also contributes to an improvement of the peak to Compton
ratio. Another advantage is that sum peaks are also detected in
sources with cascade transitions. For these reasons Ge(Li) well-
type detectors are a promising development for measuring low
gamma-activities in radiochemistry and in other fields.

300. Glos, Margaret Beach. SEMICONDUCTORS, SCINTILLATORS AND DATA
     ANALYSIS. Nucleonics, 22, no. 5 (May 1964), 50-53.

     Lithium-drifted germanium detectors, doped with gallium or
zinc, and providing deep depletion layers to give linear relation between gamma energy and pulse height spectrum, are described. Their response to various gamma sources is graphed.
Semiconductor properties, radiation damage and surface effects
of the detectors are discussed. Scintillator developments using
calcium iodide crystals are reported. Progress in data analysis
by combining computers and pulse-height analyzers in one system
is considered.

301. Goldsworthy, W. W. REDUCING CHARGE-SENSITIVE-AMPLIFIER SENSI-
     TIVITY TO DETECTOR CAPACITANCE VARIATIONS. Nucl. Instrum.
     Methods, 52 (1967), 343-44.

     A method for reducing the sensitivity charge sensitive pre-
amplifier to detector capacitance variations is described. This
method is directly applicable to many existing charge-sensitive
preamplifiers with only minor modifications.

302. Gonidec, J. P.; Walter, G., and Coche, A. CALCUL DES DISTRIBU-
     TIONS D'AMPLITUDE DANS MESURES DE SPECTROMETRIE GAMMA FAITES
     A L'AIDE DE DETECTEURS AU GERMANIUM. Nucl. Instrum. Methods,
     53 (1967), 185-191.

Taking advantage of the good energy resolution and linearity
of germanium lithium drift detectors, we describe a method for
the calculation of pulse height distributions observed with
these spectrometers for monoenergetic gamma rays. This discussion includes the main types of interactions: photoelectric,
single and double Compton effects for incident photons and for
radiations scattered from surrounding materials. The resulting
calculated gamma rays shapes are compared with experimental spectra. The detector considered was a coaxial device with a sensitive of 20 cm^3.

303. Gonidec, J. P. and Walter, G. GAMMA SPECTROMETRY WITH GE(LI)
     DETECTORS BETWEEN 0 and 14 MeV. Rev. Phys. Appl., 4 (June
     1969), 273-4. (In French) Also report CONF-681208-Suppl.

     From Conference on Experimental Methods in Nuclear and Par-
ticle Physics, Strasbourg, France.
     A method is proposed for energy calibration of a Ge(Li)
spectrometer between 0 and 14 MeV. The detection efficiency
variation of this detector in the same range of photon energy
was measured at the three resonances, $E\rho$=655, 922, and 2.489 keV,
in the $^{27}Al(p,\gamma)^{28}Si$ reaction.

304. Gordon, G. E.; Baedecker, P. A.; Anderson, C. F. L., and
     Dran, J. C. EXTENSIONS OF THE USE OF GE(LI) DETECTORS IN
     INSTRUMENTAL NEUTRON ACTIVATION ANALYSIS OF GEOLOGICAL SAM-
     PLES. Massachusetts Inst. of Tech., Cambridge. Aug. 28,
     1968. 9p.   CONF-681003-4; Also report MIT-905-125

     From International Conference on Modern Trends in Activation
Analysis, Gaithersburg, Md.
     The use of Ge(Li) detectors in instrumental neutron activa-
tion analysis was extended to very short-lived species and to
classes of geological samples not previously studied. Concentrations of Al, Mg, and V were determined in ultamafic rocks and
stony meteorites.

305. Gorni, S.; Hochner, G.; Nadav, E., and Zmora, H. TIMING CIRCUIT
     FOR GE(LI) DETECTORS. Nucl. Instrum. Methods, 53 (1967),
     349-351.

     A limiter for timing with Ge(Li) detectors is described.
The limiter gives an additional output proportional to the rise
time of the pulse, which is used to compensate for the walk
caused by the spread in pulse rise time.

306. Gossett, Charles. TECHNICAL PROBLEMS IN THE APPLICATION OF
     GE(LI) GAMMA-RAY DETECTORS. pp. 1-16 of "Nuclear Research
     With Low Energy Accelerators." Marion, Jerry B., ed. New
     York, Academic Press, 1967.   Also report CONF-670615

     From Symposium on Nuclear Research with Low-Energy Accelera-
     tors, College Park, Md.
     The factors which affect the principal characteristics of
     resolution, efficiency, counting rate and timing capabilities
     for each of the major elements of the single detector system in-
     cluding the detector itself, the preamplifier, shaping amplifier,
     optional post-amplification electronics including timing cir-
     cuitry, and the analog-to-digital converter are discussed.

307. Goulding, F. S. and Hansen, W. L. AN AUTOMATIC LITHIUM DRIFTING
     APPARATUS FOR SILICON AND GERMANIUM DETECTORS. IEEE Trans.
     Nucl. Sci., NS-11 (June 1964), 286-290.

     Drifting a thick lithium-drifted counter (silicon and ger-
     manium) is a time-consuming operation that frequently results in
     a poor device, owing to inadequate knowledge of progress of the
     drifting operation. The drifting apparatus described here auto-
     matically controls the temperature of the detector that is being
     drifted to maintain the leakage current at a preselected value.
     While drifting proceeds, a continuous measurement is made of the
     distance of the lithium-drifted region from the opposite face of
     the wafer. When the drifted region reaches 30 mil or less from
     the back of the wafer a meter indicates the thickness of the un-
     drifted region and, when this thickness falls below a preselected
     value, the temperature of the detector is automatically reduced
     to room temperature. The need for constant supervision of the
     drifting operation is thereby eliminated and reliance on theo-
     retical drift-rate calculations to predict the drift-through time
     is avoided.

308. Goulding, F. S. and Jarrett, Blair V. A METHOD FOR MAKING THIN-
     WINDOW DETECTORS. Lawrence Radiation Laboratory, University
     of California, Berkeley. Jan. 17, 1966. 9p.   UCRL-16480

     A method is described for producing thin-window lithium
     drifted germanium x- or gamma-ray detectors. The success of the
     process is indicated by a high yield of good detectors. Measure-
     ments show that the window-thickness is in the range of 1 micron
     or less and x-ray energy resolution figures of 1.4 keV (full
     width at half minimum) were obtained.

309. Goulding, F. S. and Lothrop, Robert P. SEMICONDUCTOR-DETECTORS
     HAVING A LITHIUM COMPENSATED SHELF REGION BETWEEN OPPOSITE
     CONDUCTIVITY TYPE REGIONS. Nov. 26, 1968. Patent to AEC.

     A thick semiconductor for detecting charged particle radia-
     tion and gamma rays, and having means for greatly reducing sur-
     face currents to thereby increase the signal to noise ratio is
     described.

310. Goulding, F. S. SEMICONDUCTOR DETECTORS -- THEIR PROPERTIES AND
     APPLICATIONS. Nucleonics, 22, no. 5 (May 1964), 54-61.

     Basic detector and system information is collected. Advan-
     tages of semiconductor over gas or scintillation detectors are
     given. Sensitive thickness and signal-to-noise ratio are dis-
     cussed. Types of detectors--silicon-diffused junction, surface
     barrier, lithium-drifted silicon and germanium--are reviewed.
     Requirements of detector amplifier systems are discussed, and
     circuit diagrams are given. Examples of the use of the various
     criteria for instrumentation choice are given for studies of
     fission fragments, high-energy reactions, alpha decay, beta
     spectrometry and conversion electrons.

311. Goulding, F. S. SEMICONDUCTOR DETECTORS FOR NUCLEAR SPECTROMETRY.
     Lawrence Radiation Laboratory, University of California,
     Berkeley. July 30, 1965. 178p.   UCRL 16231

     Four included lectures constitute a collection of essential
     information on semiconductor detectors and associated systems.
     The lectures concern: the descriptive theory of semiconductors,
     semiconductor technology and manufacture, factors determining
     the energy resolution in measurements with semiconductor detec-
     tors, and miscellaneous detector topics. Very few experimental
     results are given.

312. Goulding, F. S. A SURVEY OF THE APPLICATION AND LIMITATIONS OF
     VARIOUS TYPES OF DETECTORS IN RADIATION ENERGY MEASUREMENT.
     IEEE Trans. Nucl. Sci., NS-11 (June 3, 1964), 177-190.

     This paper provides a collection of data relevant to the use
     of various types of semiconductor detector in nuclear spectros-
     copy and gives examples of the use of detectors in specific ex-
     perimental applications. Basic data on absorption of various
     kinds of radiation in germanium and silicon are given, and these
     are related to the characteristics of different types of semi-
     conductor detectors.

313. Graham, R. L. TIMING CHARACTERISTICS OF LARGE COAXIAL GE(LI)
     DETECTORS FOR COINCIDENCE EXPERIMENTS. IEEE Nucl. Sci. Sym-
     posium, San Francisco, Calif. Oct. 18-20, 1965.
                                              Also report AECL-2505

     The results are presented of the study of the variation with
     position of the charge collection time and its influence on co-
     incidence resolving time. With two Ge(Li) spectrometers both de-
     tecting 511 keV events, the coincidence resolution curve has a
     full width at half-maximum of ~15 nsec and approximately expo-
     nential tails with half slopes ~5 nsec. An example of the ^{156}Gd
     level scheme is described.

314. Graudinya, L. Ya; Kalnin, J., and Pelekis, L. FULL REGISTRATION
     EFFICIENCY FOR GAMMA RAYS BY GE SEMICONDUCTOR DETECTOR.
     Latv. PSR Zinat, Akad. Vestis, Fiz. Teh. Zinat, no. 5 (1967),
     81-6. (In Russian)

     The full registration efficiency for gamma-rays by means of
     a Ge semi-conductor was calculated. The calculation is carried
     out for energies ranging from 0.1 to 10 MeV in case of cylin-
     drical detector with a radius ranging from 0.25 to 2.5 cm and
     thickness from 0.1 to 1.0 cm for various distances from the
     source to the detector.

315. Greenwood, R. C.; Helmer, R. G., and Gehrke, R. J. PRECISE
     COMPARISON AND MEASUREMENT OF GAMMA-RAY ENERGIES WITH A
     GE(LI) DETECTOR I. 50 TO 420 KEV. Nucl. Instrum. Methods,
     77 (1970), 141-58.

     A Ge(Li) gamma-ray spectrometer was used to make precise
     comparisons of the reported energies of a number of gamma-rays
     that are useful for energy calibration. The results indicate
     that in the energy region studies, below 420 keV, this spectrom-
     eter is capable of a relative precision of about 5 ev.

316. Grenier, G.; Nierat, G.; Poussier, C.; Pigneret, J., and
     Samueli, J. J. COMPENSATOR FOR SHIFTING OF TIME SPECTRA OB-
     TAINED WITH GE(LI) DETECTORS FOR ENERGIES BETWEEN 0.4 AND
     10 MEV. Nucl. Instrum. Methods, 75 (1969), 240-4. (In
     French)

     An electronic circuit which compensates the broadening of
     the time spectrum obtained with a Ge(Li) detector is described.
     An example of application to the n-gamma discrimination in a
     time-of-flight experiment is given.

317. Grenier, G. and Poussier, C. GAMMA SPECTROMETER USING A GE(LI)
DETECTOR AND OPERATING SIMULTANEOUSLY AS A PAIR SPECTROMETER
AND IN THE ANTI-COMPTON REGIME. (SPECTROMETRE GAMMA UTILISANT
UN DETECTEUR GE(LI) ET FONCTIONNANT SIMULTANEMENT EN SPECTRO-
METRE DE PAIRES ET EN-ANTI-COMPTON.) Commissariat a l'Energie
Atomique, Limeil-Brevannes,(France). Sept. 1968. 24p.
(In French)   CEA-R-3562

A description is given of a spectrometer using a Ge(Li) detector which can operate simultaneously as a pair spectrometer
and in the anti-Compton regime. The performances are given and
compared with those of other similar equipment.

318. Griffiths, R. CALCULATED TOTAL EFFICIENCIES OF COAXIAL GE(LI)
DETECTORS. Nucl. Instrum. Methods, 91 (1971), 377-9.

A program was written to compute the total efficiencies of
coaxial Ge(Li) detectors. Graphs are presented that show how
the efficiency of four typical detectors of active volumes 20,
30, 50, and 80 cm^3 varies for gamma rays of energies 0.01 to
10.0 MeV for source-to-detector distances in the range 1.0 to
25.0 cm.

319. Gruhn, C. R.; Kuo, T.; Maggiore, C.; Preedom, B.; Samuelson, L.,
and Chandler, J. LITHIUM-DRIFTED GERMANIUM FOR CHARGED PAR-
TICLE SPECTROSCOPY. IEEE Trans. Nucl. Sci., NS-15, no. 3
(June 1968), 337-346.

320. Gruhn, C. R.; Kane, J. V.; Kelly, W. H.; Kuo, T., and Berzins, G.
A SINGLE CRYSTAL GE(LI) COMPTON SPECTROMETER. Nucl. Instrum.
Methods, 54 (1967), 268-276.

A single crystal Ge(Li) Compton spectrometer has been tested
showing considerable promise as a device for detecting gamma-
rays in the presence of strong Compton and neutron backgrounds.

321. Gruhn, C. R.; Todd, R. R.; Maggiore, C. J., and Kelly, W. H.
A SINGLE CRYSTAL GE(LI) CONVERSION-COEFFICIENT SPECTROMETER.
Nucl. Instrum. Methods, 75 (1969), 109-12.

The design and operation of a conversion-coefficient spectrometer in which a single small Ge(Li) crystal serves as the detector both for gamma-rays and conversion electrons are discussed.
A calibration curve for the detector is presented, and both its

advantages and faults as compared with a spectrometer that uses
separate detectors for gamma-rays and electrons are discussed.

322. Gryksa, P.; Riehs, P., and Rumpold, K.   HIGH ENERGY GAMMA RAYS OF
     ^{76}AS FOLLOWING THERMAL AND EPITHERMAL NEUTRON CAPTURE.   Oes-
     terreichische Studiengesellschaft fuer Atomenergie G. M. b. H.
     Siebersdorf. 1969. 14p.   SGAE-PH-80/1969

     Using a Ge(Li) detector, the energies and intensities of the
     high energy gamma transitions of the reaction ^{75}As(N, gamma) ^{76}As
     have been measured. By comparison between gamma ray intensities
     in thermal and epithermal neutron capture, 6 intense transitions
     are found to be due mainly to direct neutron capture. A classi-
     fication of gamma ray intensities after epithermal neutron cap-
     ture indicated an El multipolarity for 8 gamma transitions. A
     determination of the binding energy, using the ^{36}Cl gamma energies
     for calibration gave 7300±4 keV.

323. A GUIDE TO THE USE OF GE(LI) DETECTORS.   Princeton, New Jersey;
     Princeton Gamma Tech. n.d.

324. Guinn, V. P.; Graber, F. M., and Fleishmann, D. M.   GE(LI) GAMMA-
     RAY SPECTROMETRY AS A PILOT FOR NAI(TL) GAMMA RAY SPECTROMETRY
     Talanta, 15 (Nov. 1968), 1159-63.

     Lithium-drifted germanium semiconductor detectors give much
     better resolution than do thallium-activated sodium iodide detec-
     tors, but much lower sensitivity. They can often advantageously
     be used in conjunction with NaI(Tl) detectors to show whether cor-
     rections must be applied for activities other than that measured
     and to provide the necessary information for calculation of cor-
     rections.

325. Gunnersen, E. M.   RECENT APPLICATIONS OF SEMICONDUCTOR TECHNIQUES
     IN THE STUDY OF NUCLEAR RADIATIONS.   Rep. Progr. Phys., 30
     (1967), 27-95.

     The chief topic discussed is the application of semiconductor
     counters to the analysis of the properties of nuclear radiations,
     not only in nuclear physics, but also in space physics and medi-
     cine. The criteria governing the choice of suitable semiconductor
     materials for counters, the factors limiting resolution (e.g., the
     role of trapping in large lithium-drifted counters), the transmu-
     tation doping method of compensation, the ion implantation tech-
     nique of making thin contacts, and irradiation damage effects in
     counters are discussed. Particular attention is given to the

development and applications of lithium drifted germanium, position-sensitive, and avalanche multiplication counters. Particle channelling and the potentialities of the ion implantation technique for making electronic devices to the study of basic irradiation damage mechanisms in semiconductor materials are also described.

326. Gunnink, Ray; Levy, H. B., and Niday, J. B. IDENTIFICATION AND DETERMINATION OF GAMMA EMITTERS BY COMPUTER ANALYSIS OF GE(LI) SPECTRA. Presented at the Symp. on Nucl. Methods of Chem. Analysis, Miami Beach, April 10-14, 1967.
UCID-15-140; CONF-670401-10

An application of the Ge(Li) gamma detector is its usefulness as a nondestructive analytical tool. However, the data from such systems are typically contained in 1000 to 4000 channels of a pulse height analyzer and the gamma spectra are frequently complex. To reduce the effort of spectral interpretation to a minimum, a computer code was developed that will make an identification and quantitative determination of the isotopic components producing the complex Ge(Li) gamma spectrum. In addition to the general approach and operating considerations used in a quantitative analysis of this type, features of the computer code are discussed and typical results shown. The characteristics of the computer code are location of peaks in a spectrum, determination of precise peak energies, calculation of peak intensities, identification of the isotopes responsible for the peaks by using a library tape containing the necessary decay scheme information, and final calculation and tabulation of results.

327. Gunnink, Ray and Niday, J. B. QUANTITATIVE ANALYSIS OF UNKNOWN MIXTURES BY COMPUTER REDUCTION OF GE(LI) SPECTRA. pp. 1244-5 of "Modern Trends in Activation Analysis." Vol. 2. DeVoe, James R., ed. Washington, D. C., National Bureau of Standards, 1969.
CONF-681003-Vol. 2

From International Conference on Modern Trends in Activation Analysis, Gaithersburg, Md.
Computer programs capable of accurate reduction and interpretation of gamma spectra of mixtures from a Ge(Li) detector in use at Lawrence Radiation Laboratory were modified for the treatment of unresolved peak interferences in the final calculations of the disintegration rates of various components. A brief summary of the data reduction sequence is given.

328. Gurfinkel, Y. and Notea, A. USE OF ^{133}BA AS A CALIBRATION STANDARD FOR GE(LI) DETECTORS. Nucl. Instrum. Methods, 57 (1967),

173-174.

The energies and intensities of the gamma and x-rays in the decay of ^{133}Ba were measured. This nucleus is recommended for use as a calibration standard for Ge(Li) detectors.

329. Guzzi, G.; Pauly, J.; Girardi, F., and Dorpema, B. COMPUTER PROGRAM FOR ACTIVATION ANALYSIS WITH GERMANIUM LITHIUM DRIFTED DETECTORS. European Atomic Energy Community, Ispra, Italy, Chemistry Department. Brussels. May 1967. 43p.

EUR-3469

A computer program for activation analysis problems is presented. This program is based on the measurements of photopeak surfaces of complex gamma spectra obtained by means of Ge(Li) detectors. Possible interferences from radioisotopes are evaluated, but not corrected. The program has been used routinely for the determination of radioelements in biological specimens.

330. Hall, R. N. and Soltys, T. J. HIGH PURITY GERMANIUM FOR DETECTOR FABRICATION. IEEE Trans. Nucl. Sci., N.S.-18, no. 1 (Feb. 1971), 160-165.

The preparation of large (100) oriented germanium crystals with net concentrations of electrically active impurities in the range of $1-10 \times 10^{10}$ cm^3 is described. This material has been used successfully for the fabrication of semiconductor detectors having depletion layers up to 8 mm thick, without using lithium compensation.

331. Hall, R. N.; Baertsch, R. D., and Soltys, T. J. HIGH-PURITY GERMANIUM FOR GAMMA DETECTORS. May 1968. 7p.

S-68-1088; Ar-1; NYO-3870-1

Experimental equipment and procedures for purifying germanium for use in p-i-n junction gamma detectors are described. The crystal growing furnace is described, and procedures for controlling impurities are discussed including crucible, furnace atmosphere, chemical processing, dust, source germanium, and intrinsic defects. Also preparation of p-i-n detectors is outlined.

332. Haller, William A.; Filby, Royston; Rancitelli, Louis A., and Cooper, John A. THE INSTRUMENTAL DETERMINATION OF FIFTEEN ELEMENTS IN PLANT TISSUE BY NEUTRON ACTIVATION ANALYSIS. pp. 177-183 of "Modern Trends in Activation Analysis." Vol. 1. June 1969.

CONF-681003-V.1

Described is a method for relating trace element concentrations to environmental or biological parameters by neutron activation analysis. Plant tissue-samples were irradiated and allowed to decay before gamma-ray analysis. A solid state Ge(Li) spectrometer was then used to measure short-lived activation products at various time intervals as well as after certain time long-lived activation products. Neutron activation standards were counted at appropriate intervals and compared with samples in order to determine the decay characteristics of the neutron induced radionuclides.

333. Hammerer, K. and Rauch, H. CHARGE COLLECTION GE(LI)-DETECTORS IN HIGH MAGNETIC FIELDS. Atomkernergie, 15 (1970), 281-4.

The mechanism of charge collection in semiconductor radiation detectors is altered by an external magnetic field. This leads to a variation of the electron and hole tracks and therefore the charge carrier mobility decreases. This results in a deterioration of the charge collection efficiency and an increase of the pulse risetimes. A Ge(Li)-detector was examined in the field of a superconducting magnet at field strength up to 50 kG and at various detector bias voltages. The influence on the distribution of energy and pulse rise-times was measured.

334. Hanle, H.; Stelzer, K.; Koriath, M., and Krueger, H. GE(LI)-NAI(TL) PAIR SPECTROMETER. Frankfurt Univ., West Germany. Institut fuer Kernphysik. Sept. 1970. 37p. (In German)
IKF-26

By the use of a pair spectrometer the complex spectra of high-energy γ-rays can be simplified appreciably. The pair spectrometer described in this report consists of a small Ge(Li) detector and a NaI(Tl)-annulus which is divided into four equal sectors. The annulus has a length of 12.5 cm and an outer diameter of 16.5 cm. The performance of this spectrometer has been investigated with sources of ^{24}Na and ^{226}Ra in different combinations of the sectors.

335. Hanna, G. L.; Walker, D. G., and Beach, P. M. FULL-ENERGY-PEAK EFFICIENCIES OF THREE GAMMA RAY DETECTORS. AEC Research Est., Lucas Height, (Australia). April 1967. 19p.

Two sodium iodide scintillators and a lithium drifted germanium detector are used in the analysis by gamma-ray spectrometry of gaseous fission products obtained in sweep-capsule fission product release experiments. A description is given of

the full-energy-peak efficiencies of the three detectors for the
source geometries used in counting the fission samples.

336. Hansen, Niels J. CHARGE COLLECTION AND CHARGE-COLLECTION TIME
     IN SEMI-CONDUCTOR PARTICLE DETECTORS. Argonne National Lab-
     oratory. June 1966. 27p.   ANL-7226

     Charge collection in semiconductor junction and lithium-
drifted charge-particle detectors was investigated. Analytical
solutions were obtained for the charge collected as a function
of time and for the charge collection times in terms of a parameter $B_0 = X_0/D$, the reduced range of the incident ionizing
particle. It is assumed that the stopping power dE/dx, of the
detector is constant, that there is no loss of carriers by
trapping or recombination, and that the mobilities of the carriers are constant, independent of the field strength. The impli-
cations of these simplifying assumptions are discussed. Figures
are presented to illustrate the dependence of the time pulse
shape on the parameter B_0, and optimum operating conditions for
the detectors are established. The effect on the collection time
of the high fields in the ionization column is estimated.

337. Hansen, W. L.; Pehl, R. H.; Rivet, E. J., and Goulding, F. S.
     SELECTION OF GERMANIUM FOR LITHIUM-DRIFTED RADIATION DETEC-
     TORS BY OBSERVATION OF ETCH-PIT DISTRIBUTIONS. Nucl.
     Instrum. Methods, 80 (1970), 181-6.

     This note establishes the value of observation of etch-pit
distributions as a guide in the selection of germanium for use
in lithium-drifted germanium detectors. Results are presented
showing a very good correlation between the etch-pit distribution
and detector performance for a number of crystals pulled on the
111 axis. It is inferred that mechanical strain produced by
thermal conditions in the crystal growing process is the major
source of potential charge trapping sites in the final detector.

338. Hansen, W. L. and Jarrett, B. V. TECHNIQUES FOR THE FABRICATION
     OF LITHIUM GERMANIUM GAMMA DETECTORS. Nucl. Instrum. Meth-
     ods, 31 (1964), 301-306.

     Many new problems are encountered in attempting to apply
silicon lithium drifting technology to germanium. Lithium is
much more mobile in germanium and its equilibrium solubility at
room temperature can be less than the acceptor concentration.
The low intrinsic temperature requires drifting at low temperatures and large currents and this is frustrated by the lower
thermal conductivity of germanium. Large crystals of germanium

are also very fragile and are easily damaged by thermal or
mechanical shock. However, techniques developed in this laboratory have permitted the manufacture of a large number of detectors with depletion depths to 1 cm, depleted volumes to 7 cm^3 and
energy resolution on 100 keV gamma's good as 2.1 keV fwhm. These
techniques are described and some results obtained using the detectors are presented.

339. Haouat, G.; Lachkar, J., and Sigaud, J. STUDY AND CONSTRUCTION
        OF A MULTIMODE GAMMA SPECTROMETER. RC. Accad. Naz. Lincei
        (Italy), 45, no. 5 (Nov. 1968), 281-4.

     Describes a 'multimode' gamma ray spectrometer, which permits
the separate and simultaneous detection of photons according to
three modes: total absorption photoelectric and anti-Compton
effects, pair effect. The spectrometer consists of a central
Ge(Li) detector and of a large NaI annulus split in four optically separated sectors.

340. Haouat, G.; Lachkar, J., and Sigaud, J. STUDY OF TIME RESOLUTION
        IN GERMANIUM (LI) COAXIAL DETECTORS. Commissariat a l'Energie
        Atomique, Paris. 1968. 15p. (In French)
                                        CEA-CONF-1179; CONF-680906

     From International Symposium on Nuclear Electronics, Ver-
sailles, France.
     A study was made to determine the quality of the time infor-
mation that a large volume Ge(Li) coaxial detector open at both
ends could provide. The distribution function of the pulses was
determined. This makes possible the calculation of the mean value
of the collection time and its fluctuations. The influences of
the location of the detection, of the direction of the scattered
electron, of the discrimination threshold, the polarization
voltage, and the counting rate on the time definition are considered. The theoretical predictions are in good agreement with
the experimental results. Some criteria of good operation are
derived and justified.

341. Haouat, G.; Lachkar, J., and Sigaud, J. RESEARCH AND DEVELOPMENT
        OF A MULTIMODE GAMMA SPECTROMETER. Commissariat a l'Energie
        Atomique, Brugeres le-Chatel, (France). Centre d'Etudes.
        n. d. 7p. (In French)   CEA-CONF-1200

     A multimode gamma ray spectrometer which permits the separate
and simultaneous detection of photons according to three modes:
total absorption, photoelectric and anti-Compton effects, is

described. The spectrometer consists of a central Ge(Li) detector
and of a large NaI annulus split in four optically separated sectors.

342. Harpster, J. W. CONTROL OF LITHIUM DRIFT CURRENT IN SILICON AND
     GERMANIUM. Nucl. Instrum. Methods, 48 (1967), 175-176.

343. Harry, R. J. S. GE(LI) RADIATION DETECTORS. Atoomenergie Haar
     Toepass., 12 (May 1970), 147-53, 156-64. (In Dutch)

     The resolution of detectors based on NaI(Tl), Si(Tl), and
Ge(Li) crystals and the effect of impurities on their performance
are examined. The detectors are cooled with liquid nitrogen to
prevent redistribution within the crystal. The influence of the
photo-electric, and Compton effects and pair formation on the
process of detection is examined. The excellent resolution linearity and stability of Ge(Li) detectors make it imperative to
use only the best electronic circuits and components, such as
preamplifier, amplifier and pulse-height analyzer systems. The
effect of collimation, incidence and source radiation on the
formation of the spectrum is graphically illustrated. The decay
and the distribution of the decay products of the HFR fuel elements are measured by means of a homogeneous detector, a pinhole
camera, with a small image with respect to the diameter of the
detector. The high efficiency and excellent resolution have made
it necessary to use Ge(Li) detectors for activation analyses and
for the study of n, γ reactions.

344. Harvey, J. R. FORMULA FOR PREDICTING THE SENSITIVITY OF GE(LI)
     SPECTROMETERS TO GAMMA RAYS IN THE RANGE 0.4 TO 1.5 MEV.
     Nucl. Instrum. Methods, 86 (1970), 189-97.

     A formula is developed which gives the full energy peak effi-
ciency of a Ge(Li) spectrometry system to gamma rays. Irradiation
geometry and source size and shape can be varied over wide limits.
The predictions of the formula agree well with experimental data
obtained using detectors with volumes between 0.6 cm^3 and 4 cm^3
exposed to gamma sources with photon energies in the range 0.4-
1.5 MeV.

345. Hayashi, I.; Kern, H. E.; Rodgers, J. W., and Wheatly, G. H.
     VACUUM ENCAPSULATED LITHIUM-DRIFT DETECTOR TELESCOPE. Bell
     Telephone Laboratory, Murray Hill, N. J. n. d. 26p.

     Fabrication techniques for producing lithium-drifted tele-
scopes for space experiments were developed. High resistance

grooves were obtained in lithium-drifted silicon detectors which can separate the detector into independent regions having small cross-talk as well as sharp spatial definition. This indicates the possibility of making integrated arrays of lithium-drifted detectors in a single substrate. The origin of leakage current and noise is not yet clear. The surface conditions of the n-i junctions can be a contributing factor. In addition, there are indications that the surface barrier plays a prominent role in this problem. Results of environmental life tests show that the methods of detector fabrication and vacuum encapsulation are capable of providing detectors with the stability required for space experiments.

346. Heath, R. L. DETECTOR EFFICIENCIES: A CLARIFICATION OF TERMS. pp. 252-7 of "Semiconductor Nuclear-Particle Detectors and Circuits." Brown, W. L., ed. Washington, D. C., National Acadamy of Sciences, 1969.   Also report CONF-670520

From Conference on Semiconductor Nuclear Particle Detectors and Circuits, Gatlinburg, Tenn.
The problems of finding a unique definition of efficiency for semiconductor detectors is discussed. The question of what measurements best describe the detector performance is examined.

347. Heath, R. L. ENERGY DEPENDENCE OF GAMMA-RAY LINEWIDTHS IN GERMANIUM (LITHIUM) DETECTORS. pp. 247-51 of "Semiconductor Nuclear-Particle Detectors and Circuits." Brown, W. L., ed. Washington, D. C., National Academy of Sciences, 1969.
                              Also report CONF-670520

From Conference on Semiconductor Nuclear Particle Detectors and Circuits, Gatlinburg, Tenn.
Some data on the energy dependence of gamma-ray line-widths measured by Ge(Li) detectors are presented. The phenomenon of peak broadening is discussed and a model for explaining these effects is proposed.

348. Heath, R. L.; Black, W. W., and Cline, J. E. INSTRUMENTAL REQUIREMENTS FOR HIGH RESOLUTION GAMMA RAY SPECTROMETRY USING LITHIUM-DRIFTED GERMANIUM DETECTORS. IEEE Trans. Nucl. Sci., NS-13, no. 3 (June 1966), 445-456.

The effective use of the lithium-drifted germanium gamma-ray detectors in achieving energy resolutions less than one kilovolt (FWHM) in preamplifiers.

349. Heath, R. L.   PROBLEMS IN THE CALIBRATION AND USE OF HIGH-RESO-
     LUTION GERMANIUM (LITHIUM) SPECTROMETER SYSTEMS. pp. 642-
     659 of "Semiconductor Nuclear Particle Detectors and Cir-
     cuits." Proceedings. Brown, W. L., ed. Washington, D. C.,
     National Academy of Sciences, 1969.

                                              Also report CONF-670520

         From Conference on Semiconductor Nuclear Particle Detectors
     and Circuits, Gatlinburg, Tenn.
         Limitations on measurements in nuclear gamma spectroscopy
     imposed by instrumentation are discussed, problems encountered
     in the calibration of high resolution Ge(Li) spectrometers and
     the importance of the performance criteria for each component
     of the electronic systems are examined in detail.

350. Heath, R. L.   POTENTIAL OF THE GE(LI) GAMMA-RAY SPECTROMETER FOR
     PRECISE GAMMA-RAY ENERGY MEASUREMENTS. pp. 251-77 of "Pro-
     ceedings of the Third International Conference on Atomic
     Masses, University of Manitoba." Winnipeg, Canada, University
     of Manitoba Press, 1967.   CONF-670818

         A brief description is presented of the development of the
     lithium-ion drifted germanium gamma-ray spectrometer and its po-
     tential for the precision measurement of gamma-ray energies is
     described.

351. Heath, R. L. and Johnson, L. O.   SYSTEM REQUIREMENTS FOR HIGH-
     RESOLUTION GAMMA-RAY SPECTROMETRY AT HIGH COUNTING RATES.
     pp. 141-8 of "Ispra Nuclear Electronics Symposium, Pro-
     ceedings." May 1969.   CONF-690515; Also report EUR-4289

         The full potential of high-resolution gamma-ray spectrometers
     has been severely hampered by degradation in the quality of the
     pulse-height spectrum at high input data rates. Over the past
     year or two considerable effort has been expended in the examina-
     tion of this problem. The nature of these instrumental difficul-
     ties, realistic experimental requirements, and state-of-the-art
     solutions to these problems are reviewed. A de-coupled system
     which has been developed for high-rate application is described
     and the performance presented.

352. Helmer, R. G.; Heath, R. L.; Putnam, M., and Gipson, D. H.
     PHOTOPEAK ANALYSIS PROGRAM FOR PHOTON ENERGY AND INTENSITY
     DETERMINATIONS, GE(LI) AND NAI(TL) SPECTROMETERS. Nucl.
     Instrum. Methods, 57 (Dec. 1967), 46-57.

A description is given of a method of analyzing gamma-ray spectra measured with lithium-ion drifted germanium and NaI(Tl) multichannel spectrometers. This method is based on a computer program which determines the parameters (peak position and area) of the Gaussian function which best fits the data points in a particular peak. This program includes corrections for the deviation from linearity of the electronics system and the efficiency of the detector system. Some of the peaks can have energies assigned for calibration purposes. Gamma-ray energies and intensities can then be computed for all peaks in a spectrum. From the quality of the fit to the calibration lines and the uncertainty in each energy is computed.

353. Helleboid, Jean Marie. POSITION-SENSITIVE DETECTORS FOR CHARGED PARTICLES (LES DETECTEURS A LOCALISATION POUR PARTICLES CHARGES). Commissariat a l'Energie Atomique, Grenoble, (France). Laboratoire de Physique Nucleaire. Feb. 1968. 25p. (In French)   CEA-BIB-102

In this report, we describe the possibilities of using position sensitive detectors for charged particles. The circuits associated with several detectors forming a position-sensitive set of detectors is reviewed, and the performances as well as the limits of actual realizations are discussed.

354. Henck, R.; Siffert, P., and Coche, A. CHARACTERISTICS OF A 85 cm^3 GE(LI) GAMMA-RAY DETECTOR. Nucl. Instrum. Methods, 60 (Apr. 1968), 343-5.

Characteristics and performance as a gamma-ray spectrometer of a large volume (85 cm^3) two open-ended coaxial Ge(Li) diode are described. By using a low noise field-effect tetrode transistor preamplifier a resolution value of 4.3 keV has been achieved for ^{60}Co (1332 keV). With a ^{60}CO and ^{24}Na source, the sumpeaks at 2506 keV and 4122 keV have been observed.

355. Henck, R.; Stab, L.; Lopes da Silva, G.; Siffert, P., and Coche, A. DRIFT RATE AND PRECIPITATION OF LITHIUM IN GERMANIUM. IEEE Trans. Nucl. Sci., NS-13, no. 3 (1966), 245-251.

During the study of P-I-N Ge(Li) junctions to be used for gamma ray spectrometers, it appeared that the drift process of Li+ ions is strongly depending on the nature of the raw materials, in similar conditions, the change from less than 1 mm to more than 5 mm. The influence on drift rate of several factors is studied; resistivity, nature of acceptors, (Ga, In, Zn) lifetime of minority carriers, copper diffusion.

356. Henck, R.; Siffert, P.; Miehe, J., and Coche, A. EFFICIENCY AND
     TIME RESOLUTION OF 120 cm^3 GE(LI) DETECTOR. Nucl. Instrum.
     Methods, 74 (1969), 169-70.

     The characteristics (resolution, full-energy peak efficiency,
     pulse rate times) of a 120 cm^3 Ge(Li) gamma-ray detector are
     described. The time resolution obtained at 0.511 and 1.33 MeV
     was respectively 8 and 5.4 nsec.

357. Henck, R.; Siffert, P., and Coche, A. GE(LI) TOTAL ABSORPTION
     SPECTROMETERS. Rev. Phys. Appl., 4 (June 1969), 277-8.
     (In French)   Also report CONF-681208

     From Conference on Experimental Methods in Nuclear and Parti-
     cle Physics, Strasbourg, France.
     Several two crystal Ge(Li) detector systems have been devel-
     oped for use as total absorption spectrometers, as well as a
     52 cm^3 well-type detector.

358. Henck, R.; Siffert, P.; Gutknecht, D.; De Laet, L., and
     Schoenmaekers, W. TRAPPING EFFECTS IN GE(LI) DETECTORS AND
     SEARCH FOR A CORRELATION WITH CHARACTERISTICS MEASURED ON
     THE P-TYPE CRYSTALS. IEEE Trans. Nucl. Sci., NS-17, no. 3
     (June 1970), 149-59.   Also report CONF-700301

     From 12th Scintillation and Semiconductor Counter Symposium,
     Washington, D. C.
     Various Ge(Li) detectors showing trapping effects (Low energy
     tailing for an applied field of 10^{3}V/cm) with uniform gamma-rays
     irradiation have been studied using a collimated beam of 137 Cs
     gamma-rays. The following parameters have been determined:
     velocity, mobility, drift length, and carrier lifetime. For
     many samples, a good agreement between experimental measurements
     of charge collection efficiency with the position of irradiation
     and calculated curves is obtained demonstrating that the trap
     distribution is uniform. The carrier velocities and the mobili-
     ties are practically not modified in the samples showing trapping
     effects, but the drift lengths are lowered and, as a consequence,
     the carrier lifetimes. Various measurements including infrared
     transmission spectra, etchpit distribution, electron lifetimes
     have been made on the starting P-type crystals. Results are
     presented showing a valuable correlation between the electron
     lifetimes measured at 77° K by the pulse charge method and detec-
     tor performance.

359. Hick, H. A COMPARATIVE STUDY OF SEMI-CONDUCTOR GAMMA SPECTRO-
     METERS FOR SCANNING OF IRRADIATED DRAGON FUEL ELEMENTS.
     Nucl. Instrum. Methods, 40 (1966), 337-347.

     Three types of gamma spectrometers using a Ge(Li) detector
     (Full energy peak spectrometer, anti-Compton spectrometer, Comp-
     ton-spectrometer) are compared with respect to their usefullness
     in analysis of complex spectra between 100 keV and 3 MeV. Approx-
     imate response functions for various energies are given.

360. Hick, H. and Pepelnik, R. A NEW GE(LI)-SPECTROMETER FOR THE
     DETERMINATION OF BURN-UP OF REACTOR FUEL ELEMENTS. Paper 5
     of the "Proceedings of the Meeting on the Applications of
     Ge(Li) Detectors in Science, Technology, Medicine and Indus-
     try." Brussels, 1967. 1967.   CONF-671078; Also BLG-425

     The design and operation of an improved summing lithium-
     drifted germanium Compton spectrometer are briefly discussed.

361. Hick, H. and Pepelnik, R. SUMMING GE(LI)-COMPTON SPECTROMETER
     WITH HIGH PEAK-TO-TAIL RATIO. Nucl. Instrum. Methods, 68
     (Feb. 1969), 240-4.

     The operation and performance of a summing Ge(Li)-Ge(Li)-
     Compton spectrometer are described. The geometry and the elec-
     tronic adjustment have been optimized with respect to best
     response function. Peak-to-tail ratios of 400 for 662 keV and
     360 for 1120 keV energy have been obtained. The construction of
     a summing Si(Li)-Ge(Li)-Compton spectrometer is suggested to
     avoid residual contributions of the pair production to the
     response function.

362. Higuchi, H.; Tomura, K.; Takahashi, H.; Onuma, N., and
     Hamaguchi, H. USE OF A GE(LI) DETECTOR AFTER SIMPLE CHEMICAL
     GROUP SEPARATION IN THE ACTIVATION ANALYSIS OF ROCK SAMPLES.
     PART 4: SIMULTANEOUS DETERMINATION OF STRONTIUM AND BARIUM.
     pp. 334-338 of "Modern Trends in Activation Analysis."
     Vol. 1. Washington, D. C., National Bureau of Standards,
     1969.

     Described is the rapid simultaneous determination of stron-
     tium and barium in a number of rock samples by using Ge(Li) de-
     tectors after short sample irradiation and fast chemical group
     separation by conventional precipitation methods. Strontium and
     barium abundances found in six rocks are summarized in table
     form.

363. Hill, M. W.   AN ANTICOINCIDENCE-SHIELDED GE(LI) GAMMA-RAY SPEC-
     TROMETER. <u>Nucl. Instrum. Methods</u>, 36 (Oct. 1965), 350-2.

     A lithium drifted germanium detector is used at the center
of a large plastic scintillation detector that acts as an anticoincidence shield and as a gamma spectrometer. The performance
of the detector with and without the anticoincidence shield is
shown. The shield reduces the Compton background in the measured
spectra.

364. Hiller, S.   GAMMA-SPECTROMETRIC BURNUP DETERMINATION IN REACTOR
     FUEL RODS. <u>Kerntechnik</u>, 12 (Nov. 1970), 485-90. (In Ger-
     man and English)

     The description of an experimental arrangement for the non-
destructive γ-spectrometric determination of burnup in reactor
fuel rods is followed by a discussion of the possibilities of
use of scintillation counters and of Ge(Li) diodes. The scintillation counter is suitable for measuring the relative burnup
profiles of fuel rods, whereas the Ge(Li) diode is used for absolute burnup determinations. After the arrangement was cali-
brated with fuel rod shaped standard sources, the burnup at one
point of one fuel rod was determined redundantly via the Cs^{137},
Ce^{144} and Zr^{95} gamma intensities and was compared with values
determined radiochemically. The migration of Cs^{137} was investigated.

365. Hoffer, Paul B.; Beck, Robert N.; Lembares, Nicholas; Charles-
     ton, Donald B., and Gottschalk, Alexander. USE OF LITHIUM-
     DRIFTED GERMANIUM DETECTORS FOR CLINICAL RADIONUCLIDE SCAN-
     NING. <u>J. Nucl. Med.</u>, 12 (Jan. 1971), 25-7.

366. Hofker, W. K.   DETECTION OF RADIOACTIVE RADIATION WITH P-I-N
     TRANSITIONS IN SILICON AND GERMANIUM. <u>Ned. Tijdschr. Na-
     tuur.</u>, 32 (Jan. 1966), 5-11. (In Dutch)

     The Li ion drift method for the preparation of silicon and
germanium detectors is outlined. The performance of silicon and
germanium detectors is illustrated.

367. Hollander, Jack M.   THE IMPACT OF SEMICONDUCTOR DETECTORS ON
     GAMMA-RAY AND ELECTRON SPECTROSCOPY. Lawrence Radiation
     Laboratory, University of California, Berkeley. Aug. 1,
     1965. 13p.   UCRL-16307; CONF-650832

From International Meeting on Nucl. Instrumentation, Herceg-
Novi, Yugoslavia.
A review of the semiconductor impact on gamma spectroscopy is
presented. In particular, the operating characteristics of
Ge(Li) detectors are discussed along with its many uses.

368. Hollander, Jack M. THE USE OF SEMICONDUCTOR DETECTORS FOR IN-
     TERNAL-CONVERSION COEFFICIENT MEASUREMENTS. Lawrence Radia-
     tion Laboratory, University of California, Berkeley. May
     1965. 9p.   UCRL-16035; CONF-650525-8

Presented at the International Conference on the International
Conversion Process, Nashville, Tenn.
The application of silicon and germanium detectors for measurement of internal-conversion coefficients is discussed. Data
taken by the semiconductor method are used to print out the capabilities of this new method. Some advantages over previous
methods are discussed.

369. Holm, Dale M. and Sanders, William Mort. A SEMICONDUCTOR ANTI-
     COINCIDENCE DETECTOR SYSTEM. Los Alamos Scientific Labora-
     tory, University of California, N. M. n. d. 10p.
                                                        LA-DC-6367

From Symposium on Radiochemical Methods of Analysis, Salz-
burg, Austria.
An anticoincidence lithium-drifted germanium detector system
is described. A block diagram of the circuitry is included.
Features and application of the detector are also discussed.

370. Honzatko, J. and Kajfosz, J. LINEAR POLARIZATION MEASUREMENTS BY
     GE(LI) DETECTOR. Ceskoslovenska Akademie, Ved, Rez. Ustav
     Jaderneho Vyzkumu. Dec. 1968. 20p.   UJV-2113-F

A thin planar Ge(Li) detector was used as a coincidenceless
polarimeter for measuring linear polarization of cascade gamma
quanta. The results of the measurement with ^{60}Co are presented
and compared with the theoretical curves of the polarization
sensitivity obtained by analytical calculations and by the Monte
Carlo method.

371. Hooton, B. W. THE PREPARATION OF LITHIUM-DRIFTED GERMANIUM DIODES
     AND THEIR USE AS GAMMA RAY DETECTORS. Atomic Energy Res. Est.,
     Harwell, England. July 1965. 15p.   AERE-R-4921

The advantages of Ge over Si for gamma spectroscopy are described. The operation principle and preparation stages of Li-drifted Ge reverse biased p-i-n junctions are described. During the evaluation of detector characteristics, the effective detector volume, resolution and gamma ray spectra characteristics were determined.

372. Horowitz, Y. S. and Sherman, N. K. GERMANIUM TOTAL ABSORPTION SPECTROMETER FOR 100-MeV PROTON SCATTERING. Can. J. Phys., 45 (Oct. 1967), 3265-8.

A lithium-drifted germanium total absorption spectrometer has been constructed and used to achieve an energy resolution of 0.49± 0.005 MeV in proton-scattering experiments at 100 MeV.

373. Hors, Michel and Philis, Claude. GERMANIUM JUNCTION DETECTORS. THEORETICAL AND PRACTICAL FACTORS GOVERNING THEIR USE IN RADIATION SPECTROMETRY. (DETECTEURS A JONCTION AU GERMANIUM. ELEMENTS THEORIQUES ET PRATIQUES POUR L'UTILISATION EN SPECTROMETRIE DE RAYONNEMENTS.) Nov. 1967. 89p.   CEA-R-3251

Semiconductor detectors have greatly increased the possibilities available to nuclear spectroscopist for the study of α, β and gamma radiations. Their use in radiochemistry has encouraged the study of their principle, their mechanism and also the conditions under which they can be used. Theoretical data are summarized on the best use of junction detectors, in particular Ge(Li) detectors. The laboratory work carried out over a period of one year on Ge(Li) detectors is discussed. Stress is laid on the possibilities presented by the use of these detectors as photoelectric spectrometers, and also on the precautions required. Among the numerous results presented, the resolution of 2.52 keV obtained for the gamma radiation of 145.5 keV for [141]Ce may be particularly noted.

374. Hossenlopp, J.; Haberer, A., and Bilwes, R. CHAMBRE DE REACTIONS A DETECTEURS A SEMI-CONDUCTEURS ADAPTEE A DES MESURES DE CORRELATIONS DE PARTICULES. Nucl. Instrum. Methods, 42 (1966), 137.

A rapid description of this reaction chamber with emphasis on the original elements is given. These special characteristics are: 1. three movable detectors with continuously variable solid-angle; 2. target storage box in vacuum; 3. short-lived radioactive source locks; 4. beam centering system.

375. Hotz, Henry P.; Mathiesen, J. M., and Hurley, J. P. CALCULATED
        EFFICIENCIES OF LITHIUM-DRIFTED GERMANIUM DETECTORS. Nucl.
        Instrum. Methods, 37 (Nov. 1965), 93-7.

        Gamma-ray detection efficiencies for Ge(Li) detectors were
        calculated on an IBM 704 computer for crystals of 1.0 to 10.0 mm
        and source to crystal distances for 1.0 to 25.0 cm were selected.

376. Hotz, Henry P.; Mathiesen, J. M., and Hurley, J. P. MEASUREMENT
        OF POSITRON ANNIHILATION INTERACTIONS WITH A GE(LI) DETECTOR.
        Naval Radiological Defense Laboratory, San Francisco, Calif.
        Oct. 1967. 29p.   USNRDL-TR-67-127; AD-663113

        It was observed that the annihilation radiation photopeaks in
        a Ge(Li) detector is considerably broader than that of a gamma-
        ray of the same energy. It seems reasonable to assume that the
        increased width is the result of the Doppler shift of the anni-
        hilation photopeak, i.e. the longitudinal Doppler shift of the
        radiations is measured, while the transverse shift is measured in
        the usual angular correlation experiments. By using a computer
        stripping program to remove the distortion produced by the finite
        energy resolution of the detector, momentum distributions are ob-
        tained which are in agreement with those which have been pub-
        lished. Only one detector is necessary for these measurements
        and all momenta channels are detected at once. However, detector
        energy resolution severely limits the momentum resolution. Argu-
        ments are presented which indicate that no very large improvement
        in Ge(Li) detector resolution can be expected.

377. Howes, J. H. and Knill, G. CHARACTERISTICS OF LARGE-AREA ION-
        IMPLANTED p-n JUNCTIONS FOR NUCLEAR RADIATION DETECTORS.
        pp. 97-101 of "Ion Implantation." Stevenage, Eng., Peter
        Peregrinus, Ltd., 1970.   CONF-700945

        From Conference on Ion Implantation, Reading, England.
        A description is given of Li-compensated semiconductor de-
        tectors for nuclear spectroscopy, in which the thin entry window
        is fabricated by ion doping to produce uniform non-injecting
        contacts in p-type Si and Ge crystals.

378. Hrabovcova, Alojzia. APPLICATION OF COMPUTERS IN THE ANALYSIS
        OF GAMMA SPECTRA. Jad. Energ., 16 (Jan. 1970), 37-41.
        (In Czech)

        A review of the principles of several analytical methods for
        computer analysis of complex gamma spectra obtained from NaI/Tl-
        scintillation and Ge(Li)-semiconductor spectrometers is presented.

379. Hrastnik, B.; Ljubicic, A.; Vojnovic, B.; Ilakova, K., and
     Jurcevic, M. A. GE(LI)-NAI(TL) SYSTEM FOR GAMMA-GAMMA ANULAR
     CORRELATION MEASUREMENTS. Fizika Zagreb, 1 (1969), 127-36.

   A Ge(Li)-NaI(Tl) system for coincident and directional gamma
gamma angular correlation measurements is described. The system
was checked by measuring the lifetime of the 0.482 MeV level in
^{181}Ta by the delayed coincidence method and by measuring the
anisotropy for the 1.172-1.332 MeV cascade gamma transition in
^{60}Ni. The results are in very good agreement with previous
measurements.

380. Huang, F. C.; Osman, C. H., and Ophel, T. R. LINE SHAPE ANALYSIS
     OF SPECTRA OBTAINED WITH GE(LI) DETECTORS. Australian Na-
     tional Univ., Canberra, Research School of Physical Sciences.
     1968. 19p.

   Line shape analysis techniques have been applied to gamma
ray spectra obtained with a Ge(Li) detector. It was found that
least squares fitting procedures enabled more accurate and rapid
extraction of relative intensities than was possible from the
areas of the full energy or double escape peaks.

381. Hurley, John P.; Mathiesen, James M., and Da Gragnano, Victor L.
     THE FABRICATION AND USE OF THE HIGH RESOLUTION LITHIUM-
     DRIFTED GERMANIUM GAMMA-RAY DETECTOR. July 31, 1967. 33p.
                              USN-RDL-Tr-67-55; AD-655624

    Experiments designed to study the behavior of various Ge(Li)
detector systems are described. Techniques of improving energy
resolution are discussed. Measurements of detector response to
thermal radiation are described. A series of measurements showing the variations of live width as a function of energy is discussed. Finally, the appearance of anomolous peaks in the pulse
height spectra is described.

382. Hutch, Gerald C.; McKinney, Russell, and Locker, Robert J.
     DEVELOPMENT OF A GERMANIUM AVALANCHE-TYPE SEMICONDUCTOR NU-
     CLEAR PARTICLE DETECTOR AND DISCUSSION OF AVALANCHE DETECTOR
     ARRAYS. Trans. Nucl. Sci., NS-15, no. 3 (June 1968), 246-57.
                              Also report CONF-680207

   From 11th Scintillation and Semiconductor Counter Symposium,
Washington, D. C.
   Research on avalanche or high field types of semiconductor
detectors has increased within the past two years, impelled, in

part, by the great interest in the photodetection field. The
surface contoured-deep junction diffusion approach was concen-
trated on while other investigators generally have pursued the
use of a planar-oxide passivated-shallow diffused structure. Reported are results of studies of deep (25-50 microns) diffused
n+ p type germanium avalanche detector structures. A relatively
simple closed box type process has been developed for diffusion
of arsenic and antimony into germanium. Results of detector
characteristics and multiplication measurements are presented.
A concept for arrays of avalanche detectors (particularly to this
type of silicon) is discussed. Results, in terms of uniformity
of avalanche breakdown and gain of a number of arrays fabricated
are indicated together with the insight that this work has given
to what appears to be the limitation of avalanche devices--
resistivity inhomogeneities in the starting semiconductor crystals. These inhomogeneities take the form, at least in crystal
produced by the floating zone technique, of micro-striations.
An example of a developed array is shown. It is a large area
quadrature instrument that, it is hoped, will find use in detection of plutonium in wounds by detecting its 17 keV x-ray
emanation.

383. Huth, Gerald C.; McKinney, Russell A., and Locker, Robert J.
     DEVELOPMENT OF A GERMANIUM AVALANCHE-TYPE SEMICONDUCTOR NU-
     CLEAR PARTICLE DETECTOR AND DISCUSSION OF AVALANCHE DETECTOR
     ARRAYS. General Electric Co., Philadelphia, Penn. IEEE
     Trans. Nucl. Sci., NS-15, no. 3 (June 1968), 246-257.

384. Imhof, W. L. HIGH-RESOLUTION GAMMA-RAY SPECTROSCOPY STUDY.
     Lockheed Missiles and Space Co., Sunnydale, Calif. Dec. 31,
     1969. 21p.   LMSC-695167

     The basic objective of the program is to study methods by
which nuclear debris measurements can be made more extensive and
interpretable through utilization of important new advances in
high-resolution gamma ray spectroscopy. The program focussed on
two principal areas; (1) a laboratory program to investigate,
for selected weapons materials, the fission product gamma-ray
spectra obtained under various conditions and (2) a study of the
utility and feasibility of using Ge(Li) detectors on satellites
to obtain diagnostic information on nuclear weapons detonated
in the atmosphere or at high altitudes. The preliminary survey
laboratory measurements of the gamma rays emitted from neutron
induced fission fragments have been completed. On the basis of
the preliminary laboratory measurements, estimates have been
made of the responses expected in space and a proposed experiment
has evolved for conducting a proper evaluation of the operational

model. These various activities are briefly reviewed in the report.

385. Ingebretsen, F. CALCULATION OF DOUBLE ESCAPE PEAK ANGULAR COR-
      RELATION ATTENUATION COEFFICIENTS FOR SQUARE GE(LI) GAMMA
      DETECTORS. Nucl. Instrum. Methods, 60 (Apr. 1968), 308-16.

     Angular correlation attenuation coefficients for the double
escape peak are calculated for nine Ge(Li) detectors, by the use
of a simple Monte Carlo technique. The calculations are valid
for photon energies between 1.5-6 MeV. The statistical model
used is empirically adjusted to fit the observed intrinsic efficiency for energies from 6-10 MeV. This gives an approximation
to the attenuation coefficients for these energies. The coefficients are compared to the "any interaction" attenuation coefficients.

386. Inoue, Tamon. ANALYSIS FOR GAMMA RADIATION SPECTRUM BY COMPUTER.
      Bunseki Kiki, 6 (Sept. 1968), 616-19. (In Japanese)

     A method for automatic analysis by computer of complex gamma
spectra detected with a high resolution Ge(Li) detector was
described and applied to a few examples.

387. Inouye, Tamon. A METHOD FOR THE ANALYSIS OF COMPLEX GAMMA-RAY
      SPECTRA USING A COMPUTER. "Ispra Nuclear Electronics Sym-
      posium." Proceedings. May 1969. 1969. 416p.
                                          CONF-690515; EUR-4289

     A computer method is developed for the analysis of complex
gamma-ray spectra obtained by the Ge(Li) detector. The computer
code by this method carries out smoothing, background subtraction
and peak sorting of the multichannel pulse height distribution.
By applying this method, 800 channel pulse height distributions
of gamma-ray spectra have been analyzed.

388. INSTRUMENTATION AND ELECTRONICS. pp. 185-212 of "Research Labo-
      ratories. Annual Report for the Period, Jan. - Dec. 1966."
      Israel Atomic Energy Comm., Tel Aviv. July 1967.   IA-1128

     Various types of electronic equipment and measuring instru-
ments are discussed. Some of these include beta spectrometers,
semiconductor detectors, pulse analyzers, reactor pneumatic
systems, electrometers, temperature control equipment, power
supplies, and equipment for studying sealing mechanisms of
vacuum gasket seals.

389. INSTRUMENTATION TECHNIQUES. pp. 163-95 of "Radium and Meso-
thorium Poisoning and Dosimetry and Instrumentation Techniques in Applied Radioactivity." Annual Progress Report.
Mass. Inst. Tech. May 1964.

Data are presented on the design, calibration, and performance of a large volume scintillation counter for Rn counting; an automated α-γ correlation system designed to investigate nuclear properties in the trans lead region, a small vacuum
system for the preparation of thin-film evaporated α sources,
Li-drifted Ge semiconductors for Rn counter applications, a
α scanning system using new crystal suspensions, and a α-ray-
focusing collimator.

390. Irigaray, J. L. and Petit, G. Y. STUDY OF THE CHARACTERISTICS
OF A 120 cm^3 GE(LI) DETECTOR AND COMPARISON WITH SMALLER
DETECTOR. Nucl. Instrum. Methods, 80 (1970), 764-7. (In
French)

391. Ishaq, A. F. M.; Khan, H. A.; Najam, M. R.; Khan, Z. U.; Qazi,
M. N.; Ahmad, A. A. Z., and Ahmed, N. M. EFFICIENCY CALIBRA-
TION OF A GE(LI) DETECTOR. J. Natur. Sci. Math., 8 (Oct.
1968), 165-9.

Attention is called to the usefulness of the well-estab-
lished beta-gamma coincidence technique for low-energy efficiency
calibration of Ge(Li) detectors. Results of the high-energy efficiency calibration using neutron capture gamma rays are also
presented.

392. Ishizuka, Y.; Kurano, Y.; Yoshii, N., and Nagahara, Y. FABRICA-
TION OF A LARGE VOLUME GE(LI) DETECTOR. Sci. Rep. Yokohama
Nat. Univ. 1(Japan), no. 15 (Jan. 1969), 9-16.

The fabrication of a large volume Ge(Li) detector of single
open-ended type is reported. The material loss was minimized
compared with Fiedler's method. Furthermore, the heat-up process was interposed during the drift procedure, which distinctly
improved the drift efficiency. The volume of the final detector is 35 cc. and the energy resolution is 5.2 keV at 2.754 MeV
gamma-ray from ^{24}Na.

393. Ishü, Mitsuhiko; Tsutomu, Tamura; Naomota, Shikozano, and
Takekoshi, Hidekuni. FABRICATION OF THICK GERMANIUM DETEC-
TORS. Japan Atomic Energy Res. Inst., Tokyo, Japan.
Nov. 1966. 12p.   JAERI-1131

Lithium-drifted germanium detectors of depletion depths up to 8 mm and diameter about 19 mm were fabricated from pulled indium-doped, p-type single crystals. The detectors have an energy resolution of 4-5 keV (FWHM) for gamma rays from ^{60}Co, when used with a conventional low-noise preamplifier. The procedure for fabricating the thick germanium detectors is described in detail.

394. Israel, H. I.; Lier, D. W., and Storm, E. COMPARISON OF DETEC-
     TORS USED IN MEASUREMENT OF 10 to 300 KEV X-RAY SPECTRA.
     Nucl. Instrum. Methods, 91 (1971), 141-57.

X-ray spectral measurements in the energy range 10-300 keV were made using four detectors (sodium iodide crystal, germanium, and silicon semiconductors, and xenon proportional counter). The measured spectra are compared. The resolution of each detector was measured, and its effect on the continuum and on the characteristic L and K lines of the target is discussed. The results of Monte Carlo calculations of photopeak efficiency, escape-peak losses, and the Compton distribution for each detector are presented. The distortion of incident monoenergetic lines is shown for each type of detector. The region of usefulness of each detector is discussed, and an iterative method which utilizes all four types of detectors to deduce the true incident spectrum is described.

395. Iyengar, K. V. L. and Lal, B. CALCULATED EFFICIENCIES OF
     CYLINDRICAL GE(LI) DETECTORS. Indian J. Phys., 42 (Feb.
     1968), 85-90.

Detection efficiencies of Ge(Li) detectors of cylindrical geometry for point sources placed on the axis of the detector were calculated for gamma ray energies ranging from 0.01 to 10.00 MeV on a CDC 3600 computer. Detectors of area of cross section 2 to 10 sq. cm. and of various depletion depths were considered for source to crystal distances ranging from 1-25 cm.

396. Jackson, C. N., Jr. FFTF STATE-OF-THE-ART REPORT FOR LOW-LEVEL
     NEUTRON FLUX INSTRUMENTATION. Battelle-Northwest, Rich-
     land, Wash., Pacific Northwest Lab. Feb. 1969. 21p.
                                                          BNWL-946

397. Jackson, Glen L. PREPARATION OF PLANAR LITHIUM-DRIFTED GERMANIUM
     DETECTORS. Ames Lab., Iowa. Feb. 1969. 62p. Thesis.
                                            Also report IS-T-296

Lithium has been drifted into germanium using both boiling-liquid and plate-type drift apparatus. Both methods have been investigated at two temperatures, the 320.8°K boiling temperature of Freon 113 and the 297°K boiling temperature of Freon 11. It was found that the lithium drift rate was closer to the theoretical value at the higher temperature than it was at the lower temperature. The drift rate should be closer to the theoretical value at the higher temperature because of the solubility of lithium in germanium rises with temperature. In one plate-type drifter the temperature and drift voltage were maintained constant, but the drift current varied. In the other plate-type drifter the drift voltage and drift current were maintained constant, but the temperature varied. Both plate-type drifters allowed lithium to be drifted into germanium to a depth of 5 mm, but the leakage current of 20 mA for the constant-voltage, constant-current process was much less than the leakage current for the constant-temperature, constant-voltage process.

398. Jackson, H. E.; Julien, J.; Samour, C.; Chevillon, P. L.; Morgenstern, J., and Netter, F. RESONANCE NEUTRON CAPTURE WITH LITHIUM-DRIFTED GERMANIUM DETECTORS. pp. 154-162 of "Panel on Lithium-Drifted Germanium Detectors." Proceedings. Vienna, International Atomic Energy Agency, 1966.

399. Jaklevic, J. M.; Bernthal, F. M.; Radeloff, J. O., and Thompson, M. G. A PULSE-HEIGHT COMPENSATION SYSTEM FOR GE(LI) TIMING. Nucl. Instrum. Methods, 69 (Mar. 1969), 109-14.

A simple pulse-height compensation system for correcting the energy walk to a Ge(Li) fast-timing data is described. The results of the method applied to a number of different Ge(Li) detectors are presented. A factor-of-two improvement in time resolution was found to be possible in all cases.

400. Jamini, M. A. PRODUCTION OF LITHIUM DRIFTED GERMANIUM DETECTORS BY A C DRIFT. Brookhaven National Laboratory, Upton, N. Y. 1966. 17p.   BNL-10683; CONF-661020-2

Presented at 13th Annual Nuclear Science Symposium, Boston. Thick planar detectors were produced by a c drifting. Lithium is diffused on two opposite surfaces of a p-germanium block, thus producing an n+ pn+ structure with two diodes back to back. Drift the proceeds from the two surfaces on alternate half cycles of the ac field. When the two compensated regions meet, one n-contact is removed and replaced by a p-contact. Any remaining uncompensated regions are then compensated by dc drift. The p-contact is produced by diffusing gallium-indium entectic at 400°C.

401. Janarek, F. J.; Helenberg, H. W., and Mann, H. M. GOLD PLATING PROCEDURE FOR A THIN WINDOW ON GERMANIUM GAMMA RAY SPECTROMETER. Rev. Sci. Instrum., 36 (1966), 1501-1502.

402. Jenkins, David A. A TECHNIQUE FOR DRIFTING LITHIUM INTO GERMANIUM. University of California, Berkeley, Lawrence Radiation Laboratory, Inorganic Materials Research Div. Jan. 1967. 12p.   UCRL-17317

    A method that was used successfully to drift Li into Ge is described. A large variation in the performance of samples taken from different ingots is found. For a given ingot the same results were obtained with different samples, so fluctuations in performance were attributed to the original ingot and not to variations in the techniques of fabrication. The lithium drift mechanism is also very sensitive to the presence of impurities.

403. Jewell, R. W.; John, W.; Massey, R., and Saunders, B. G. BENT CRYSTAL SPECTROMETER FOR USE WITH A GERMANIUM DETECTOR. Nucl. Instrum. Methods, 62 (1968), 68-76.

    A bent crystal spectrometer has been constructed in the Cauchois geometry for use with a Ge detector. The design is mechanically simple and capable of high accuracy. In the usual slit scanning mode of operation the Ge detector improves the signal-to-background ratio. Another mode is described utilizing the bent crystals to preselect an energy band for analysis by the Ge detector, thus suppressing Compton interference. A new measurement has been made of the .59 keV gamma ray from 241 am with the result: $E\gamma = 59.536 \pm 0.001$ keV.

404. Jindal, G. R. and Faust, J. W., Jr. MATERIALS PROBLEMS IN LITHIUM-DRIFTED GERMANIUM RADIATION DETECTORS. J. Appl. Phys., 41 (1970), 2106-9.

    This paper describes some materials problems involved in the preparation of lithium-drifted germanium gamma-ray detectors. A preliminary study shows some decided differences between the characteristics of the original starting material and the characteristics of the material after fabrication procedures.

405. John, Joseph; Orphan, V. J., and Hoot, C. G. NEARLY MONOENERGETIC
     SOURCE OF 6.129 MeV GAMMA RAYS FOR GE(LI) DETECTOR CALIBRA-
     TION. Nucl. Instrum. Methods, 75 (1969), 271-3.

     A simple method of producing a source of high energy gamma
rays is described. The gamma rays are obtained from the decay
of ^{16}N nuclei that have been produced by the reaction ^{16}O(n,p)
^{16}N. The 6.129 MeV gamma ray is completely free of distortion,
Doppler broadening, and energy shift. Except for the presence
of about 7% of 7.117 MeV gamma rays, this source is essentially
monochromatic and provides an excellent means of studying the
line shape of a Ge(Li) detector.

406. Johns Hopkins University. Dept. of Physics. NUCLEAR REACTIONS
     IN LIGHT NUCLEI AND MOESSBAUER STUDIES. Oct. 1969. 144p.
                                                        NYO-2028-4

     The design and performance of Compton lithium drifted ger-
manium for linear polarization measurements are described.

407. Jonasson, L. G. LOW NOISE SOLID STATE AMPLIFIERS FOR SEMICON-
     DUCTOR DETECTORS. Nucl. Instrum. Methods, 26 (1964),
     104-108.

     The use of a tunnel diode and of a field effect transistor
in the input stage of an amplifier for semiconductor detectors
is discussed.

408. Jones, D. W. NEW CIRCUIT FOR PULSE SHAPE DISCRIMINATION.
     Trans. Nucl. Sci., NS-15, no. 3 (June 1968), 491-9.
                                        Also report CONF-680207

     From 11th Scintillation and Semiconductor Counter Symposium,
Washington, D. C.
     A new electronic circuit has been developed for discriminat-
ing between pulses induced by neutrons and gammas in organic
detectors. Both stilbene and NE 213 may be used as detectors
with no change in the pulse shape discriminator circuit.

409. Kalbitzer, S. FABRICATION AND USE OF LITHIUM-DRIFTED GERMANIUM
     DETECTORS. pp. 133-141 of "Panel on Lithium-Drifted Ger-
     manium Detectors." Proceedings. Vienna, International
     Atomic Energy Agency. 1966.

     The preparation of Ge(Li) detectors by several techniques

is described, including pulse drift, d. c. drift and a. c. drift.
Results of drifting planar, radial and multiple structures are
reported.

410. Kalbitzer, S. PROCESS FOR THE MANUFACTURING OF VERY HIGH
        QUALITY RADIATION DETECTORS. May 1969. French Patent
        1,570,841. (In French)

411. Kane, W. R. and Mariscotti, M. A. AN EMPIRICAL METHOD FOR
        DETERMINING THE RELATIVE EFFICIENCY OF A GE(LI) GAMMA-RAY
        DETECTOR. Nucl. Instrum. Methods, 56 (1967), 189-196.

        An empirical method for determining the full-energy peak
        efficiency between 0.2 and 3.0 MeV and the two-escape peak
        efficiency between 2 and 9 MeV of a Ge(Li) gamma-ray detector
        is described. The full-energy peak efficiency is determined
        with the use of several nuclides which emit gamma rays with ac-
        curately known intensity ratios and a linear least-square fit-
        ting procedure.

412. Kantele, J. and Suominen, P. A GE(LI)-GE(LI) SUM-PEAK (SUMMING
        COINCIDENCE) SPECTROMETER. Nucl. Instrum. Methods, 86, no. 1
        (Sept. 1970), 65-76.

413. Kantele, J. and Suominen, P. A GE(LI) GAMMA SPECTROMETER EM-
        PLOYING AN ANTI-COMPTON MANTLE OF NAI(TL). Nucl. Instrum.
        Methods, 41 (1966), 41-4.

        An anti-Compton spectrometer employing a Ge(Li) detector sur-
        rounded by a large (20 cm by 30 cm) NaI(Tl) annulus is described.
        The system also includes a smaller annular crystal of NaI(Tl)
        which detects degraded gammas scattered from the germanium de-
        tector in the backward direction. The cryostat and the vacuum
        chamber of the spectrometer are constructed to give a large re-
        duction of the Compton tails of gamma spectra when the system
        is operated in the anti-Compton mode. The reduction of the tail
        of the ^{60}Co spectrum near the Compton edge of the 1.17 MeV gamma
        is approximately 16 and the average reduction in the region above
        0.3 MeV is nearly 10. The use of the spectrometer configuration
        is also proposed.

414. Kantele, J. and Suominen, P. A SIMPLE SUMMING COMPTON GE(LI)
        SPECTROMETER. Nucl. Instrum. Methods, 56 (1967), 351-354.

        A summing Compton spectrometer employing two Ge(Li) detectors

is described. One of the crystals detects the Compton electron
and the other the secondary quantum scattered through a means
angle of about 135°.

415. Kaskey, Edwin and Rickey, Martin. PREPARATION OF GERMANIUM DE-
     TECTORS. Rev. Sci. Instrum., 35, no. 10 (1964), 1364.

     A method of locating regions of a Li-drifted germanium de-
     tector responsible for improper l-V characteristics is described.
     After under-going lithium diffusion and etching, the detector is
     held between two spring contacts and dipped slowly into liquid
     nitrogen. It is then allowed to warm up in air with about 22V
     of reverse bias applied yielding a frost covered device. Pref-
     erential heating results, with the frost melting first in regions
     of high current. These specific areas can then be ground or cut
     off, with only a small reduction in the size of the device.

416. Katzenstein, H. S. and Ziemba, F. P. DEVELOPMENT, TESTING AND
     EVALUATION OF GERMANIUM NUCLEAR PARTICLE DETECTORS. Army
     Nucl. Defense Laboratory, Edgewood Arsenal, Md. June 1966.
     110p.   AD-634718

     A study of some of the properties of thin window germanium
     nuclear particle detectors which were fabricated by use of the
     surface barrier lithium-ion-drift technique.

417. Kaufmann, C. STANDARDIZATION OF SEMICONDUCTOR DETECTORS. pp. 66-
     69 of "Internationale Arbeitstagung Herstellung und Anwendung
     von Halbleiterdetektoren." Sept. 1963. (In German)

     Proposals are made for the standardization of semiconductor
     detectors, especially surface barrier detectors, so that the
     conditions of preparation and coupling would be the same. The
     suggestions are intended only as the basis for discussion among
     the users of semiconductor detectors and not as a rigid guideline.

418. Kaufmann, C. THE STATE OF DEVELOPMENT OF SEMICONDUCTOR DETECTORS
     IN THE CENTRAL INSTITUTE FOR NUCLEAR PHYSICS AT ROSSENDORF.
     pp. 10-15 of "Internationale Arbeitstagung und Anwendung von
     Halbleiterdetektoren, 23-28 of September, 1963 in Rossen-
     dorf." 1963. (In German)   ZFK-PHA-12

     The surface barrier detectors prepared from p-Si were de-
     scribed. The initial material and the properties of these de-
     tectors were discussed. A special ageing phenomenon was detected.

The electronics, especially the preamplifiers used, were described. Future developments to be expected were mentioned.

419. Kawada, Y.; Yura, O., and Kimura, M. RADIOACTIVITY MEASUREMENTS
     BY THE $4\pi\beta-\gamma$ ANTICOINCIDENCE SPECTROSCOPY METHOD USING A
     GE(LI) DETECTOR. Nucl. Instrum. Methods, 78 (1970), 77-85.

     By the aid of a $4\pi\beta-\gamma$ anticoincidence set-up with a Ge(Li)
     detector, a method was developed which makes possible radioac-
     tivity measurements of some nuclides having complex decay
     schemes without requiring the decay scheme correction.

420. Kegel, Gunter H. R. PRECIPITATION OF LITHIUM IN GERMANIUM DURING
     ION DRIFT. IEEE Tran. Nucl. Sci., NS-15, no. 3 (June 1968),
     332-336.

     The precipitation of lithium in germanium has been investi-
     gated by Morin and Reiss and by Swalin and Weltzin. The theo-
     ries developed by investigators have been used to calculate the
     rate of lithium precipitation in a typical lithium-drifted
     germanium detector.

421. Kemmer, J. CONSTRUCTION OF A HIGH RESOLUTION GAMMA SPECTROMETER
     AND EXAMPLES FOR ITS USE. (AUFBAU EINES HOCHAUFLOESENDEN
     GAMMA-SPEKTROMETERS UND BEISPIELE FUER SEINE ANWENDUNG).
     pp. 146-155 in "Soc. Europeene de Prot. Contre des Rayonne-
     ments Detectors in Radiation Protec. and Radiation Measure-
     ment Tech." Sept. 1967. (In German)

     A high resolution gamma spectrometer with a Li-drifted Ge
     crystal was constructed. The Ge crystal, the mechanical struc-
     ture with cryostats and shielding, and the associated electronics
     are described. As application of the use of the spectrometer, the
     gamma spectra of the lanthanides ^{154}Eu, ^{159}Gd, ^{160}Tb, ^{110}Ag and
     ^{124}(122)Sb were obtained and compared with the spectra obtained
     with a NaI(Tl) crystal.

422. Kemmer, J. EIN GE(LI)-GAMMA SPEKROMETER FUR GERINGE GAMMA-
     ACTIVITATEN. Nucl. Instrum. Methods, 64, no. 3 (1968), 268.
     (In German)

     By using a well-type Ge(Li)-diode with a drifted volume of
     9 cm^3 mounted in a special cryostat, it becomes possible to
     measure gamma-activities with 2% absolute efficiency at 500 keV.

423. Kennett, Terence J. INSTRUMENTATION FOR GAMMA-RAY SPECTROSCOPY.
     Phys. Today, 19, no. 7 (July 1966), 86-9.

     The development of semiconductor detector as computer tech-
     nology along with analog-to-digital converters for collection
     storage, and analysis of gamma spectra are reviewed.

424. Khvashchevska, Ya.; Yurkovski, Ya.; Shimchak, M., and Skochek,
     K. COAXIAL GERMANIUM DETECTORS OF GAMMA RADIATION. Joint
     Institute of Nuclear Research, Warsaw, Poland. 1969. 14p.
     (In Russian)   INR-P-1161

     Coaxial Ge(Li) detectors having active volumes of 2 to 10 cm^3
     and resolutions of 1.9 keV for 122 keV gamma radiation (^{57}Co)
     and 2.7 keV for 662 keV gamma radiation (^{137}Cs) were fabricated.
     Preamplifiers for these detectors and the vacuum chamber for the
     operation at low temperatures (N_2) have been constructed.

425. Kimerling, L. C.; Golovin, L. B., and Gatos, Harry C. GERMANIUM
     RADIATION DETECTORS COMPENSATED BY IRRADIATION DEFECTS.
     Proc. IEEE, 57 (Feb. 1969), 208-9.

     A method is described for the preparation of room-temperature-
     stable germanium radiation detectors. Some preliminary observa-
     tions are presented concerning the role of materials and fabri-
     cation parameters in detector performance.

426. Kitahara, Tanemichi; Gotoh, Hiroshi, and Shiraii, Eiji. MEAS-
     UREMENTS OF FISSION GASES IN AN IN-PILE LOOP WITH GE(LI)
     DETECTORS. J. Nucl. Sci. Technol., (Tokyo), 5 (Nov. 1968),
     596-8.

427. Klein, Claude A. SEMICONDUCTOR PARTICLE DETECTOR: A REASSESSMENT
     OF THE FANO FACTOR SITUATION. IEEE Trans. Nucl. Sci., NS-15,
     no. 3 (June 1968), 214-225.

     From 11th Scintillation and Semiconductor Counter Symposium,
     Washington, D. C.
     The problems dealt with concern the production of electron-
     hole pairs in a semiconductor nuclear radiation sensor. The
     goal is to develop a semiphenomenological model capable of
     describing the present experimental situation from the stand-
     point of yield, variance and band-gap dependence.

428. Klotz, G. and Walter, G. A STUDY OF (n, n'γ) REACTIONS WITH
     GE(LI) DETECTORS. Rev. Phys. Appl., 4 (June 1969), 271-2.
     (In French)   Also report CONF-681208

     From Conference on Experimental Methods in Nuclear and Par-
     ticle Physics, Strasbourg, France.
     A high resolution gamma-ray spectrometer using a Ge(Li)
     detector for (n, n'γ) measurements at 14.5 MeV bombarding
     energy, is described. Preliminary results for the ^{31}P(n, n'γ) ^{31}P
     reaction are reported.

429. Kopechy, J.; Ratynski, W., and Warming, E. CURVES FOR THE
     RESPONSE OF A GE(LI) DETECTOR TO GAMMA RAYS IN THE ENERGY
     RANGE UP TO 11 MeV. Nucl. Instrum. Methods, 50 (1967),
     333-339.

     This paper discusses the response function of a Ge(Li) co-
     axial detector with a sensitive volume of 17 cm^3 for gamma rays
     of energies ranging from 2.23 to 10.83 MeV.

430. Korf, H. and Schmand, J. METHOD FOR CHEMICAL SURFACE TREATMENT
     OF GE(LI) P-I-N DETECTORS. Nucl. Instrum. Methods, 72
     (1969), 353-4.

     A reproducible method for chemically forming the surfaces of
     Ge(Li) detectors is described. Most of the Ge detectors were
     well suited for γ-spectroscopy.

431. Kosina, Z. PEAK FINDING METHOD FOR USE IN GE(LI) SPECTRA PRO-
     CESSING. Nucl. Instrum. Methods, 88 (1970), 163-4.

     The present method is based on elimination of background by
     subtracting a smoothed spectrum from the original one. As a
     result it provides the first guess values of peak positions and
     intensities for a conventional least-squares-fit procedure.

432. Kotina, I. M.; Meshcherskaya, L. I.; Novikov, S. R.; Ryvkin,
     S. M., and Khusainov, A. Kh. GAMMA SPECTROMETRY ON THE BASIS
     OF GERMANIUM IRRADIATED WITH NEUTRONS. Fiz. Tekh. Poluprov,
     3 (Apr. 1969), 632-5. (In Russian)

433. Koyama, Masaki and Yuki, Makoto. THE HALF-COLUMN TYPE PIN DIODE.
     J. Nucl. Sci. Technol. (Tokyo), 5 (Jan. 1968), 18-21.

The techniques of preparation and the performance characteristics obtained with the half-column type diode are described.

434. Kraner, H. W.; Chasman, C., and Jones, K. W. EFFECTS PRODUCED
     BY FAST NEUTRON BOMBARDMENT OF GE(LI) GAMMA RAY DETECTORS.
     Nucl. Instrum. Methods, 62 (1968), 173-183.

     The effects of fast neutron bombardment on Ge(Li) gamma ray
     detectors have been investigated at neutron energies of 1.10,
     5.75, and 16.49 MeV. Measurements were made of reverse leakage
     current capacitance, rise-time, energy resolution, and pulse
     height decrease as a function of neutron dose. A degradation
     of energy resolution was the most notable effect and occurred
     for each energy after approximately $10^{10} n/cm$.

435. Kraner, H. W.; Sovka, J. A., and Breckenridge, R. W., Jr. AN
     EFFICIENT DEWAR FOR LITHIUM DRIFTED GERMANIUM DETECTORS.
     Nucl. Instrum. Methods, 36 (Oct. 1965), 328-30.

     The design of a simple liquid-nitrogen dewar for use with
     Li-drifted Ge detectors is described. The "hold" time required
     for the collant to boil off is 24 or 65 hours, depending on the
     reflecting material used. The vessel uses 1.35 liter of liquid
     nitrogen.

436. Kraner, H. W. and Chase, R. L. A TOTAL ABSORPTION GE(LI) GAMMA
     RAY SPECTROMETER. Brookhaven National Laboratory, Upton,
     N. Y. 1968. 20p.   BNL-12332; CONF-680207-4

     A configuration of a large Ge(Li) detector has been developed
     which suppresses single, local interactions, particularly Compton
     scatterings. The peripheral n+ contact of a coaxial detector is
     split into two or more segments, and events are recorded from the
     central p contact if signals are coincidentally present on two
     segments of the n+ contact.

437. Kuchly, J. M.; Stab, L.; Henck, R.; Siffert, P., and Coche, A.
     MOBILITE DES ELECTRONS ET DES TROUS A 77°K DANS LE GERMANIUM
     COMPENSE AU LITHIUM. Nucl. Instrum. Methods, 47 (1967),
     148-150.

     The drift mobility of electrons and holes has been measured
     at 77°K. The observed minority carrier mobility is less than
     the mobility calculated from scattering by phonons and impurity
     atoms. The discrepancy, which is greater than a factor of 5 in

some cases, has been attributed to collisions with substitutional
oxygen (which has been measured by lithium precipitation).

438. Kuehn, E. W. and Schroeder, A. N. F.   CAPACITANCE MEASUREMENT ON
        BIASED GE(LI) AND SI(LI) RADIATION DETECTORS USED WITH
        COOLED FET INPUT. Nucl. Instrum. Methods, 79 (1970), 304-8.

    A method is described to measure the capacitance of cooled
    pin radiation detectors as a function of bias, by using the
    standard spectroscopy equipment, i.e., a charge sensitive pre-
    amplifier, pulse shaping amplifier, multichannel analyzer and a
    test pulser.

439. Kumar, Satinder and Garg, R. K.   CALCULATION OF CHARGE AND CUR-
        RENT PULSES IN PLANAR AND COAXIAL GE(LI) DETECTORS. Univ.
        of Delhi. Proc. Indian Nat. Sci. Acad. Part A. 36. (March
        1970), 101-11.

    The charge and current pulses in a planar and a coaxial
    Ge(Li) detector, due to a gamma quantum entering the detector,
    are calculated from a common theoretical approach. The pulse
    shapes and rise times of these pulses are also calculated and
    discussed for a special case in which the (e-, h)-pairs are
    produced throughout the sensitive volume of the detector.

440. Kurtz, E.; Battleson, K.; McDaniels, D. K., and Horen, D. J.
        RESONANCE ENERGY AND YIELD MEASUREMENTS WITH A GE(LI) DE-
        TECTOR. Nucl. Instrum. Methods, 69 (Mar. 1969), 56-60.

    Two simple innovations for measurement of resonance energies
    and yields have been tested using the $^{27}Al(p,\gamma)^{28}Si$ reaction.
    The first involves the use of a thick (several keV) target and
    simultaneous excitation of resonances over a broad energy range,
    while the second uses a thin (20 keV) target and single reso-
    nances are excited successively. Both methods exploit the high
    resolution of a Ge(Li) detector and the excellent linearity and
    gain stability of the associated electronics.

441. Laegsgaard, E.; Martin, F. W., and Gibson, W. M.   POSITION-SENSI-
        TIVE SEMICONDUCTOR PARTICLE DETECTORS FABRICATED BY ION
        IMPLANTATION. IEEE Trans. Nucl. Sci., NS-15, no. 3 (June
        1968), 329-345.

442. Lal, B. and Iyengar, K. V. K.   PHOTO- AND DOUBLE-ESCAPE EFFI-
        CIENCIES OF CYLINDRICAL GE(LI) DETECTORS. Bhabha Atomic

Research Centre, Bombay, (India). 1969. 24p.   BARC-403

Monte Carlo calculations have been made to compute the full
energy peak efficiencies and the double escape peak efficiencies
for right circular Ge(Li) detectors for point gamma-ray sources.
The full energy peak efficiencies are calculated in the gamma-
energy range from 100 keV to 2.5 MeV and the double escape peak
efficiencies are calculated in the range from 1.7 MeV to 4.0 MeV.
All calculations have been made for a source to detector distance of 5 cm.

443 Lal, B. and Iyengar, K. V. K. MONTE CARLO CALCULATION OF GAMMA-
       RAY RESPONSE CHARACTERISTICS OF CYLINDRICAL GE(LI) DETECTORS.
       pp. 228-33 of "Proceedings of Nucl. Phys. Solid State Phys.
       Symposium, Bombay, India." 1968.   CONF-681210-Vol. 2

From Nuclear Physics and Solid State Physics Symposium, Bombay, India.
Calculated response function characteristics including full
energy peak efficiency and double escape peak efficiency are
given for various-sized cylindrical lithium-drifted germanium
detectors. The Monte Carlo calculational method is discussed.

444. Lalovic, B. A DUAL ELEMENT COAXIAL GE(LI) GAMMA RAY SPECTROMETER.
       Nucl. Instrum. Methods, 47 (1967), 173-175.

Discusses the construction of high resolution germanium
gamma ray spectrometers with sensitive volumes up to 50 cm^3.

445. Lalovic, B.; Azuma, R. E., and Petrovic, B. LARGE MULTIPLE ELE-
       MENT GERMANIUM SPECTROMETERS. IEEE Trans. Nucl. Sci., NS-14,
       no. 1-2 (Feb. 1967), 514-520.

Large coaxial Ge(Li) diodes have been assembled in multiple
arrays in one cryostat in order to provide spectrometers of very
large sensitive volumes. A double element and a four element
spectrometer are described in detail. The latter spectrometer
has a total sensitive volume of 180 cm^3. The efficiencies and
resolutions of the spectrometers were measured for various combinations of diodes and summing procedure.

446. Lalovic, B. MULTIPLE-DETECTOR ARRAYS. pp. 230-6 of "Semicon-
       ductor Nuclear Particle Detectors and Circuits." Brown,
       W. L., ed. Washington, D. C., National Academy of Sciences,
       1969.   Also report CONF-670520

From Conference on Semiconductor Nuclear Particle Detectors
and Circuits, Gatlinburg, Tenn.

447. Lamb, J. F.; Prussin, S. G.; Harris, J. A., and Hollander, J. M.
     APPLICATION OF LITHIUM-DRIFTED GERMANIUM GAMMA-RAY DETECTORS
     TO NEUTRON ACTIVATION ANALYSIS. NON DESTRUCTIVE ANALYSIS OF
     A SULFIDE ORE. Anal. Chem., 38 (June 1966), 813-18.

     Lithium-drifted germanium gamma spectrometers were used for
     non-destructive analysis of a sulfide ore by neutron activation.
     The concentrations of ten elements were determined in the range
     from 1.8×10^5 ppM (iron) to 2.0 ppM (gold). The analytical
     sensitivity obtained with available Ge(Li) detectors was compared
     with sodium iodide scintillators for estimation of manganese in
     the ore. The various detector properties which affect sensitivi-
     ties are discussed.

448. Larsen, R. N. and Strauss, M. G. SEPARATELY HOUSED GE(LI) DE-
     TECTORS OPERATING IN PARALLEL AND AS A SUMMING COMPTON SPEC-
     TROMETER. IEEE Trans. Nucl. Sci., NS-17, no.3 (June 1970),
     254-64.   Also report CONF-700301.

     From 12th Scintillation and Semiconductor Counter Symposium,
     Washington, D. C.
     Coaxial as well as planar Ge(Li) gamma-ray detectors housed
     in separate cryostats have been operated in parallel and as
     asumming Compton spectrometer. The linear signals from two de-
     tectors were direct coupled into a single charge-sensitive pre-
     amplifier, thus eliminating the need for matching electronic
     gain and linearity of two separate channels. The full-energy
     peak efficiency of two parallel coaxial detectors is nearly
     equal to the sum of the efficiencies of the individual detectors.
     Similar increase in efficiency was also observed for double-
     escape peaks when two small planar detectors were paralleled.

449. Lauber, A. and Malmsten, B. AN EXPERIMENTAL STUDY OF THE ACCU-
     RACY OF COMPENSATION IN LITHIUM DRIFTED GERMANIUM DETECTORS.
     Aktiebolaget Atomenergi, Nykoping, Sweden. Oct. 1969. 16p.
                                                        AE-373

     The nature and magnitude of the space charge existing in the
     compensated layer of lithium drifted germanium detectors was
     studied as a function of drifted depth and of the electric field
     applied during drift. Experimental values were obtained from
     the dependence of detector capacitance on applied bias.

450. Lauber, A. and Rosencrantz, B. A NEEDLE-TYPE P-I-N JUNCTION
     SEMICONDUCTOR DETECTOR FOR IN-VIVO MEASUREMENTS OF BETA
     TRACER ACTIVITY. Aktiebolaget Atomenergi, Stockholm. 1964.
     18p.   AE-162

     A miniature detector probe was developed for in vivo detec-
     tion of beta tracer activity. A lithium-drifted p-i-n semicon-
     ductor detector shaped as a cylinder 0.9 mm in diameter and
     3 mm long acts as the sensing element. The detector is encased
     in a stainless steel tube 50 mm long, fastened to a holder fitted
     with a miniature coaxial contact. The free end of the tube has
     a syringe-like, entirely tight tip. The steel tube has an outer
     diameter of 1.4 mm corresponding to a wall thickness of 0.05 mm.
     The detector is placed in the 1.1 mm part of the tube. The con-
     struction and the properties of the probe are described.

451. Lauber, A. ON THE THEORY OF COMPENSATION IN LITHIUM-DRIFTED
     SEMICONDUCTOR DETECTORS. Nucl. Instrum. Methods, 75, no. 2
     (Nov. 15, 1969), 297-308.

     The lithium ion drift method produces detectors with a
     highly but not perfectly compensated intrinsic region. The
     amount of fixed space charge left in the compensated layer and
     its dependence on drift and clean-up parameters is of great
     practical interest. The imperfect compensation is mainly due
     to the presence in the compensated layer of thermally generated
     electron-hole pairs swept apart by the voltage applied to the
     detector during drift.

452. Lauber, A.; Malmsten, B., and Rosencrantz, E. SPECIAL CRYOSTATS
     FOR LITHIUM COMPENSATED GERMANIUM DETECTORS. Aktiebolaget
     Atomenergi, Stockholm, Sweden. May 1968. 18p.   AE-320

453. Laughlin, John S. BIOLOGICAL AND CLINICAL DOSIMETRY. Annual
     Progress Report July 1, 1969-June 30, 1970. Sloan-Kettering
     Inst. for Cancer Research, New York. 1970. 79p.
                                                      NYO-3510-15

454. Lawson, E. M.; Tavendale, A. J., and Dawson, A. C. LARGE VOLUME,
     MULTI-ELEMENT GE(LI) SPECTROMETER. Australian Atomic Energy
     Commission Research Est., Lucas Heights. Nov. 1970. 15p.
                                                      AAEC/TM-572

455. Lawson, E. M. GAMMA-RAY COMPENSATED GERMANIUM DETECTORS.

Australian Atomic Energy Commission Research Establishment,
Lucas Heights. Dec. 1969. 14p.   AAEC/TM-527

456. Lederer, C. Michael. COMPUTER ANALYSIS OF SPECTRA. University
     of California, Berkeley, Lawrence Radiation Lab., Sept. 1969.
     55p.   CONF-690818-9; Also report UCRL-18948

     From International Conference on Radioactivity in Nuclear
     Spectroscopy, Nashville, Tenn.
     Computers have contributed greatly to automation in the
     analysis of multichannel spectra and, more importantly, to the
     accuracies which can be achieved. The fundamental part of most
     spectral analysis methods -- peak-shape analysis -- is reviewed
     with emphasis on the present and potential level of accuracy in
     the determination of energies and intensities. The methods
     are illustrated with analysis of Ge(Li) gamma-ray spectra, al-
     though the techniques are quite general for high-resolution,
     multichannel spectra. It is found that gamma-ray energies can
     be determined to a few eV if suitable standards are available.

457. Lee, Y. K. and Owen, G. E. A GE(LI) COMPTON EFFECT GAMMA RAY
     POLARIZATION. Nucl. Instrum. Methods, 74 (1969), 176-8.

     The gamma ray detector which is sensitive to linear polari-
     zation of the gamma rays has been fabricated from a single
     planar Ge(Li). The operation of the detector has been tested
     by use of the 1.36 MeV gamma ray following the reaction ^{24}Mg(p,p')
     ^{24}Mg.

458. Levy, A. J.; Ritter, R. C., and Ziock, K. THE ANNULAR LITHIUM-
     DRIFTED GERMANIUM DETECTOR AS A STANDARD-GEOMETRY GAMMA
     SPECTROMETER. IEEE Trans. Nucl. Sci., NS-14, no. 1-2 (Feb.
     1967), 509-513.

     Lithium-drifted germanium detectors have been fabricated in
     the form of a cylindrical annulus. The forming of the annular
     shape was made feasible by using a spark cutting machine.

459. Levy, A. J.; Ritter, R. C., and Ziock, K. AN ANNULAR LITHIUM-
     DRIFTED GERMANIUM DETECTOR FOR STUDYING NUCLEAR REACTION
     GAMMA-RAYS. Sept. 1968. 85p.   NASA-CR-1138

     A fabrication technique was developed for truly annular Ge(Li)
     gamma detectors. Featuring spark-cutting to precisely control
     the undamaged germanium volume, the detector is capable of

standard-geometry techniques such as are used with NaI detectors.

460. Levy, A. J. and Ritter, R. C. LINEARITY OF A GE(LI) DETECTOR IN
     THE RANGE 0.5 to 10 MEV. Nucl. Instrum. Methods, 49 (1967),
     359-360.

     The linearity of an annular germanium lithium drifted gamma
     detector has been measured to be <0.046% over a range of 0.5 to
     10 MeV.

461. Lippert, J. A CRYOSTAT FOR GERMANIUM DETECTORS. Nucl. Instrum.
     Methods, 32 (1965), 360-61.

462. Lippert, J. SOME APPLICATIONS FOR SEMICONDUCTOR DETECTORS IN
     HEALTH PHYSICS. pp. 271-7 of "Radiation Protection." Part
     1. Snyder, W. S. and Abee, H. H., eds. New York, Pergamon
     Press, 1968.   Also report CONF-660920-V.1

     From 1st International Congress of the International Radia-
     tion Protection Association, Rome, Italy.
     Continuing earlier detector development, a small group in
     the Health Physics Department started work in the semiconductor
     detector field in 1963, with a feeling that these might be
     valuable for spectrometry as well as low level counting. Ger-
     manium detectors, which are produced in sizes up to 9 cm^2 at the
     time of writing (March 1966), are discussed.

463. Lipsett, J. J. and Palmer, J. F. LOCATING FUEL FAILURES BY FIS-
     SION PRODUCT DEPOSITION IN CANDU-PHW REACTORS. Atomic
     Energy of Canada Ltd., Chalk River (Ontario). Nov. 1967.
     27p.   CONF-671119-1; Also report AECL-2786

     From Conference on Failed Fuel Element Detection, Vienna,
     Austria.
     Many systems for locating fuel cladding failures are costly
     and difficult to maintain because they usually employ a complex
     arrangement of sample flows controlled in long lengths of small-
     diameter tubing. In power reactors of the pressure-tube type,
     the problem may be simplified with a Ge(Li) gamma-ray spectro-
     meter mounted near the end face of the reactor and set to scan
     the outlet coolant pipes for deposited fission products.

464. Litherland, A. E.; Alexander, T. K., and Jeffs, A. T. A SOURCE
     OF 3.85 MEV GAMMA RAYS FOR TESTING GERMANIUM (LITHIUM)

DETECTORS. pp. 684-7 of "Semiconductor Nuclear-Particle Detectors and Circuits." Brown, W. L., ed. Washington, D. C., National Academy of Sciences, 1969.

Also report CONF-670520

From Conference on Semiconductor Nuclear Particle Detectors and Circuits, Gatlinburg, Tenn.

A source of monoenergetic 3.85 MeV gamma rays and lower-energy gamma rays was produced by using $^{10}B(\alpha, p\gamma)^{13}C$ reactions occurring in an intimate mixture of ^{10}B and PuO_2. The reaction as a Q-value of 4.07 MeV, and the first three excited states at 3.09, 3.68, and 3.85 MeV in ^{13}C are populated. The 3.85 MeV level decays to the ground state giving rise to 3.85 MeV gamma ray. This gamma ray is suitable for testing Ge(Li) detectors.

465. Litherland, A. E.; Ewan, G. T., and Lam, S. T. USE OF SINGLE PLANAR GE(LI) DETECTORS AS GAMMA-RAY POLARIMETERS. Can. J. Phys., 48 (Oct. 1, 1970), 2320-30.

Also report AECL-3710

466. LITHIUM DRIFTED GERMANIUM NON-DISPERSIVE X-RAY SPECTROMETER. Cern. Cour., 3 (Mar. 1969), 92p.

Recent advances in the development of germanium x-ray spectrometers are summarized and technical data listed. The use of these devices in the field of research, medicine, industry, and non-destructive analysis is discussed, and the higher efficiency relative to silicon is emphasized.

467. Llacer, J. LITHIUM DRIFTED SEMICONDUCTOR RADIATION DETECTOR. U. S. Patent 3,413,528. Nov. 26, 1968.

A method of producing a lithium-drifted semiconductor radiation detector by cutting a groove have a U-shaped cross section defining a closed figure in one face of the detector is presented.

468. Llacer, J. WORK ON GAMMA COMPENSATED AND HIGH-PURITY GERMANIUM RADIATION DETECTORS. Summary of Report, Jan. 1-Dec. 31, 1970. Brookhaven National Lab., Upton, New York. 1970. 76p.

BNL-15643

A report of the research on the compensation of n-type germanium by gamma radiation.

469. Lombard, S. M. and Isenhour, T. L. NEUTRON CAPTURE GAMMA-RAY
     ACTIVATION ANALYSIS USING LITHIUM DRIFTED GERMANIUM SEMI-
     CONDUCTOR DETECTORS. Anal. Chem., 40 (1968), 1990-4.

     The performance of a planar and a coaxial Ge(Li) semicon-
     ductor gamma-ray detector are compared with that of a NaI(Tl)
     scintillation detector for neutron capture gamma-ray activation
     analysis. The superior resolution of these large volume Ge(Li)
     diodes more than compensates for their lower efficiency because
     post-irradiation chemical or half-life resolution is not possible
     in the capture gamma-ray method.

470. Lopes da Silva, G.; Henck, R.; Siffert, P., and Coche, A.
     LITHIUM DRIFTABILITY IN GERMANIUM. pp. 201-6 of "Semicon-
     ductor Nuclear Particle Detectors and Circuits." Brown,
     W. L., ed. Washington, D. C., National Academy of Sciences,
     1969.   Also report CONF-670520

     From Conference on Semiconductor Nuclear Particle Detectors
     and Circuits, Gatlinburg, Tenn.

471. Lopes da Silva, G.; Henck, R., and Siffert, P. PROBLEMS IN
     SELECTING GERMANIUM CRYSTALS FOR LARGE VOLUME GE(LI)
     COUNTERS. pp. 65-80 of the "Proceedings of the Meeting on
     Special Techniques and Materials for Semiconductor Detectors.
     Ispra, Italy, 1968." June 1969.   EUR-4269

     Unsolved problems encountered in fabricating and selecting
     large-volume germanium detectors for optimum performance for
     spectroscopy are discussed. These problems include irregular
     and reduced lithium drift mobility, carrier trapping effects,
     and capacitance relaxation phenomena.

472. Lowrey, A. R. and Friddell, K. D. THE RESPONSE OF P-I-N DETECTORS
     TO VARIOUS NUCLEAR RADIATION ENVIRONMENTS. Boeing Co.,
     Seattle, Washington. 1964. 50p.   D2-84071-1

     Si and Ge p-i-n detectors are exposed to various transient
     and steady-state radiation environments, and the theoretical
     and experimental variations in response are tabulated, compared
     and analyzed. The radiations to which the detectors are exposed
     include gamma rays, x-rays and electrons.

473. Ludwig, E. J. LITHIUM-DRIFTED POSITION SENSITIVE DETECTOR. Rev.
     Sci. Instrum., 36 (Aug. 1965), 1175-6.

An attempt was made to produce position sensitive detectors with depletion depths of several millimeters by lithium drifting. The technique used involves evaporating a thin metallic layer onto the side opposite the lithium diffusion to provide a uniform resistive sheet.

474. Luuko, A. and Holmberg, P. COINCIDENT SUMMING EFFECTS IN GE(LI) DETECTORS. Nucl. Instrum. Methods, 65 (Oct. 15, 1968), 121-2.

Coincident summing effects in Ge(Li) detectors in gamma spectroscopy are discussed. Consideration is given to nuclei which emit two gamma rays in cascade.

475. McConnon, D. USE OF SOLID STATE DETECTORS FOR ORGAN COUNTING. Battelle-Northwest, Richland, Washington. Pacific Northwest Laboratory. July 1970. 13p.   BNWL-1178

Results of in vivo measurement of a complex mixture of gamma emitters with conventional whole body or organ counting systems are often unsatisfactory because of the poor resolution of the counting system. The possibility of eliminating this problem by the use of a counting system employing a high resolution solid state detector prompted study of such a system to determine its suitability in measuring amounts of radionuclides in individual organs. The response of a dual crystal Ge(Li) detector to gamma emitters contained in various organ compartments of an anthropomorphic phantom was measured. The study indicated that even though its sensitivity is poor relative to that of conventional organ counting systems, use of a solid state system may yield useful supplemental information for assessment of internal depositions, provided the amount deposited in the organ is sufficiently large, and that the background and counting geometry are controlled.

476. McCready, V. R.; Parker, R. P.; Gunnersen, E. M., and others. CLINICAL TESTS ON A PROTOTYPE SEMICONDUCTOR GAMMA-CAMERA. Brit. J. Radiol., 44 (Jan. 1971), 58-62.

A prototype γ-camera was constructed from a slice of lithium drifted germanium 44 x 44 mm in area and 6.5 mm in sensitive thickness. Positional information was obtained using the orthogonal strip principle, the inherent spatial resolution being 3 mm. Data are presented on energy resolution, uniformity, linearity of image and spatial resolution. The application to imaging distributions of ^{99m}Tc in rats and humans is demonstrated, and comparisons on picture quality and background made (using the identical sources) with a conventional Anger system.

477. McDaniels, David K. and Battleson, Kirk. GERMANIUM DETECTORS AND THE COMPTON EFFECT. Am. J. Phys., 35 (Sept. 1967), 837-43.

Recent advances in solid-state lithium drifted germanium detectors have opened a new era in gamma-ray spectroscopy. A typical detector system is described which could easily be incorporated into an advanced teaching laboratory.

478. McIntyre, R. J. ON THE ORIGIN OF LEAKAGE CURRENT IN GE(LI) GAMMA-RAY DETECTORS. IEEE Trans. Nucl. Sci., NS-15, no. 5 (Oct. 1968), 6-8.

Hole-electron pair generation in Ge(Li) gamma-ray detectors due to thermal radiation from the cryostat outer casing is suggested as an often over-looked source of detector leakage current. Thermal radiation-induced currents of the order of 10^{-9} A are estimated for a typical large coaxial detector at 77°K when the cryostat outer casing is at 300°K.

479. McKee, B. T. A. TIMING OF SEMICONDUCTOR DETECTORS WITH A CONSTANT FRACTION DESCRIMINATOR. Nucl. Instrum. Methods, 62 (1968), 333-336.

A simple constant-fraction discriminator times pulses from a planar Ge(Li) detector. Time resolution is close to that with leading-edge discrimination and there is negligible time walk.

480. McKee, R. J.; Hargrove, C. K., and Smith, A. G. PRECISE CALIBRA-TION OF A GE(LI)-SPECTROMETER USING A DIGITAL TO ANALOG CON-VERTER. Nucl. Instrum. Methods, 92 (1971), 421-32.

A new method to precisely calibrate the linear electronics used with Ge(Li)-spectrometers has been developed. The calibra-tion is done by measuring the system's differential response with a mercury relay pulse generator whose amplitude is distri-buted over the range of the spectrum by a computer controlled 15 bit digital to analog converter. It is shown that the integral of the differential response, collected simultaneously with the data, linearizes the electronics throughout the whole spectrum range in a model independent way, except for small errors intro-duced by electronic noise and accidental coincidence with detec-tor events. A test of the calibration system using a small

planar detector is presented in which a linear fit of peak
position to adopted energy was done for seven γ-ray lines
between 100 and 500 keV which are known to very high precision.
The rms deviation from the linear fit for two separate runs is
less than 3 eV. In addition, a new value of 136.473 ± 0.015
keV is reported for one of the principal transitions of ^{57}Co.

481. McKenzie, J. M. and Donovan, P. F. SELECTION OF GERMANIUM
     LITHIUM DRIFT DETECTORS FOR GAMMA RAY SPECTROSCOPY. Nucl.
     Instrum. Methods, 54 (1967), 147-149.

     A few very high resolution large germanium lithium drift
detectors have been reported (<4 keV ^{60}Co, >20 cm^3). The resolution of the large detectors currently available, however, is
much poorer than these exceptional units. Smaller detectors
with high resolution are much more easily obtainable.

482. McKenzie, J. M. and deWit, R. C. SURFACE TREATMENT OF GERMANIUM
     LITHIUM DIODES. IEEE Trans. Nucl. Sci., NS-15, no. 1 (1968),
     444-447.

     Several surface treatments are described, which change the
I-V characteristics of lithium drifted germanium diodes. These
treatments, applied after the final etch, have reduced the diode
leakage current to <10-^{9}A for fields of up to 600v/mm. Treatments achieving this on several diodes are described.

483. Madden, T. C. and Gibson, W. M. UNIFORM AND STABLE dE/dx P-N
     JUNCTION PARTICLE DETECTORS. IEEE Trans. Nucl. Sci., NS-11
     (1964), 254-261.

     Solid state dE/dx E particle identification systems require
thin dE/dx detectors having a thin degree of thickness uniformity.
The lack of this uniformity has been a limiting factor in
achieving optimum results with these systems. In addition, re-
producible and straight forward fabrication techniques, long
term stability, and ruggedness are desired for the routine use
of such device.

484. Makino, M. Q.; Waddell, C. N., and Eisberg, R. M. THE NUCLEAR
     REACTION EFFICIENCY CORRECTION FOR SILICON AND GERMANIUM
     DETECTORS. Nucl. Instrum. Methods, 60 (1968), 109-112.

     The proton total reaction cross section of silicon has been
measured to be 732- 20mb at 28 - 1 MeV. The efficiency correc-

tions for nuclear protons from 5 to 150 MeV has been calculated using the energy dependence of the proton total-reaction cross sections.

485. Malm, H. L. ENCAPSULATED COAXIAL GE(LI) DETECTORS OPERATED IN
     PARALLEL. IEEE Trans. Nucl. Sci., NS-14, no. 1-2 (Feb. 1967),
     521-522.

486. Malm, H. L. IMPROVEMENTS IN LARGE VOLUME COAXIAL GERMANIUM
     SPECTROMETERS. March 1966. 31p.   AECL-2550; CRGP-1236

     From the 10th Scintillation and Semiconductor Counter Sym-
     posium, Washington, D. C.
     Large volume germanium lithium drift diodes coaxially
     drifted with one open end showed excellent performance as gamma-
     ray spectrometers. However, limitations were found in the
     coincidence resolving times possibly due to the large variations
     in charge collection times which result from the nonuniformity
     of the electric field, especially in the closed end portion.

487. Malm, H. L.; Tavendale, A. J., and Fowler, I. L. A LARGE VOLUME
     COAXIAL LITHIUM DRIFT GERMANIUM GAMMA-RAY SPECTROMETER. Can.
     J. Phys., 43 (July 1965), 1173-1181. Also report AECL-2301

     A high-resolution, germanium p-i-n diode gamma ray spectrom-
     eter was made using the coaxial method lithium drift. The de-
     tector described is 16 cm^3 in sensitive volume, three to four
     times that of the largest "planar" drifted diodes of this type
     described to date.

488. Malm, H. L. and Fowler, I. L. LARGE VOLUME COAXIAL GERMANIUM
     GAMMA-RAY SPECTROMETERS. Atomic Energy of Canada, Ltd.,
     Chalk River (Ontario). Oct. 1965. 36p.
                                     AECL-2504; CRGP-1224

     From the 12th Nuclear Science Symposium, San Francisco, 1965.
     Germanium lithium-drift p-i-n diodes for high resolution
     gamma-ray spectroscopy have been made with sensitive volumes in
     the range 16 to 54 cm^3 using the coaxial method of drift. Details
     of the construction, mounting, and preparation as spectrometers
     are given. The shape of the undrifted p-type core has been
     determined by scanning with a collimated gamma ray beam and by
     copper plating the surface of a sectioned diode. The character-
     istics as spectrometers, using gamma rays of energies in the
     range 0.1 to 7.5 MeV are shown.

489. Malm, H. L. and Fowler, I. L.   A LARGE-VOLUME HIGH-RESOLUTION
        GERMANIUM (LITHIUM) SPECTROMETER.   pp.237-40 of "Semicon-
        ductor Nuclear-Particle Detectors and Circuits."   Brown,
        W. L., ed.   Washington, D. C., National Academy of Sciences,
        1969.   Also report CONF-670520

        From the Conference on Semiconductor Nuclear Particle Detec-
        tors and Circuits, Gatlinburg, Tennessee.
        The development, design, and performance of spectrometers
        using large-volume, high resolution Ge(Li) detectors are dis-
        cussed.

490. Mann, H. M.; Bilger, H. R., and Sherman, I. S.   OBSERVATIONS ON
        THE ENERGY RESOLUTION OF GERMANIUM DETECTORS FOR 0.1 - 10
        MEV GAMMA RAYS.   IEEE Trans. Nucl. Sci., NS-13, no 3 (June
        1966), 252-264.

491. Mann, H. M.; Janarek, F. J., and Helenberg, H. W.   PREPARATION OF
        LARGE VOLUME PLANAR GERMANIUM DETECTORS AND RESULTS OF THEIR
        USE IN NUCLEAR PHYSICS EXPERIMENTS.   IEEE Trans. Nucl. Sci.,
        NS-13, no. 3 (June 1966), 336-350.

        Improved performance of germanium detectors, prepared by use
        of lithium drift techniques, was obtained by means of measurements
        of diode properties and surface resistivity during preparation.
        A compensated thickness of 16 mm was obtained for several detec-
        tors in a single drift operation.

492. Mann, H. M.   PRESENT STATUS OF THE DEVELOPMENT OF GE(LI) SPECTROM-
        ETER SYSTEMS AT ARGONNE.   pp. 118-87 of the "Proceedings of the
        Conference on Slow-Neutron Capture Gamma-Ray Spectroscopy.
        Argonne National Lab., 1966."   Throw, F. E., ed.   Nov. 1968.
                                                ANL-7282; CONF-661113

        The development of Ge(Li) spectrometers at Argonne National
        Laboratory is discussed.   Topics include the Ge(Li) detector
        including a simple physical description and dual-drift and thin-
        window detectors; cryostats; Ge(Li) spectrometer design including
        the charge-sensitive section, output amplifiers, multichannel
        analyzers, pulse generators, and present limitations and new
        areas, and details on spectrometer performance.

493. Mann, H. M.   PROGRESS IN THE APPLICATION OF SEMICONDUCTOR DETEC-
        TORS TO NUCLEAR PHYSICS EXPERIMENTS.   IEEE Trans. Nucl. Sci.,
        NS-12, no. 2 (April, 1965), 88-89.

The selection of a detector for a particular experiment is
briefly discussed, together with experiments in which phosphorus-
diffused silicon detectors and lithium-drifted germanium detectors are used. The performances and areas of applicability of
these detectors are considered.

494. Marcinkowski, A.; Rzewuski, H., and Werner, Z. RANGE-ENERGY FOR
     LOW ENERGY PROTONS IN SI AND GE. Nucl. Instrum. Methods,
     57 (1967), 338-340.

The range-energy relations for protons of energy between 0.8
MeV and 2.0 MeV in Si and Ge were determined. For the measurement of the proton energy a 100 channel pulse-height analyzer
and silicon semiconductor detector were used.

495. Marcus, Luc. ADVANCES IN THE STUDY OF GERMANIUM DETECTORS AND
     THEIR USE FOR CHARGED PARTICLES, EXPERIMENTAL STUDY OF (d,d'),
     (d, p), (d, t) REACTIONS ON ^{12}C at 80 MeV. Paris University,
     Orsay, France. 1970. 87p. (In French) Thesis.
                                             Also report NP-18570

496. Marcus, L.; Duhamel, G., and Langevin-Joliot, H. USES OF GER-
     MANIUM DETECTORS FOR CHARGED PARTICLES. Rev. Phys. Appl.,
     4 (June 1969), 279-280.   Also report CONF-681208

From the Conference on Experimental Methods in Nuclear and
Particle Physics, Strasbourg, France.
     Lithium-drift germanium detectors have been used for 80 MeV
deuterons, with energy resolution close to that of the beam.
Using side-entry technique, trapping phenomena have been seen,
similar to those observed with collimated gamma rays, which result in a peak shift. Text in French.

497. Marisotti, M. AUTOMATIC ANALYSIS OF SPECTRA FROM GE(LI) DETEC-
     TORS. pp. 233-44 of "The Conference on Slow-Neutron-Capture
     Gamma-ray Spectroscopy." Throw, F. E., ed. Nov. 1968.
                                          CONF-661113; ANL-7282

A method is described for the automatic analysis of gamma
spectra from Ge(Li) detectors. The data are recorded on magnetic
tape. A program was developed to read the tape and find the
peaks and set the required parameters to link automatically into
a least-squares fitting program.

498. Marshall, III, J. Howard; Weeks, Donald R., and Halpern, Edward.
     GAMMA-RAY SPECTROMETER FOR SPACE APPLICATIONS. pp. 98-106
     of "Nucleonics in Aerospace. Proceedings of the 2nd Inter-
     national Symposium ... Columbus, Ohio, 1967." New York,
     Plenum Press, 1968.   Also report CONF-670714

     Development of a gamma spectrometer for earth orbiting
     satellites for measuring gamma radiation emitted by astronomical
     bodies is presented. The detector system consists of a colli-
     mated array of 48 lithium-drifted germanium detectors.

499. Marten, R. KRYOSTAT FUER GE(LI)-DETEKTOREN MIT AUTOMATISCHER
     NACHFUELLUNG AUS EINEN KUEHLMITTELRESERVOIR. Nucl. Instrum.
     Methods, 57 (1967), 274-276.

     A small cryostat for cooling Ge(Li)-detectors with liquid
     N_2 is described. It is refilled through a flexible tube from a
     supply dewar by means of an automatically operating device.

500. Marten, R. and Rosomm, H. NONDESTRUCTIVE BURN UP DETERMINATION
     BY GAMMA-SPECTROSCOPY. PART 1. AN AUTOMATIC DEVICE FOR
     THE DETERMINATION OF BURN UP PROFILES IN FUEL RODS.
     Atomkernenergie, 14 (Sept.-Oct. 1969), 339-41. (In German)

     An automatically working γ-scanning device for the determina-
     tion of burn up of fuel rods is described, that permits measure-
     ments in steps down to 0.02 mm. The equipment consists of a
     scanning machine, a collimator system, and a Ge(Li)-spectrometer.

501. Martin, M. J.; Harvey, J. A., and Slaughter, G.G. GAMMA-RAY
     SPECTRA FROM RESONANCE NEUTRON CAPTURE USING A REACTOR BEAM.
     pp. 14-18 of "The Conference on Slow-Neutron-Capture Gamma-
     Ray Spectroscopy." Throw, F. E., ed. Nov. 1968.
                         ANL-7282; Also report CONF-661113

     Some measurements of gamma spectra from resonance neutron
     capture in aluminum, iron, and tin in a reactor spectrum are re-
     ported. Measurements were made with a Ge(Li) detector.

502. Martin, Robert. DEVIATION OF THE ASYMMETRY RATIO FOR A PLANAR
     GE(LI) GAMMA RAY POLARIMETER. IEEE Trans. Nucl. Sci., NS-17,
     no. 3 (June 1970), 248-53.   Also report CONF-700301

     The integral expression, which gives the probability of a
     scattered gamma ray causing a coincidence count between the

center and one of the side detectors, is derived using the
assumptions of only a Compton scattering in the center detector
and a photoelectric reaction in one of the end detectors, the
necessity of leaving at least minimum detectable amount of
energy in both detectors, and the fact that only a certain pro-
portion of the gamma rays leave the center detector with just
one scattering has not been included in this calculation.

503. Martini, M.; McMath, T. A., and Fowler, I. L. EFFECTS OF
     OPERATING TEMPERATURE ON THE BEHAVIOUR OF SEMICONDUCTOR
     DETECTORS. IEEE Trans. Nucl. Sci., NS-17, no. 3 (June 1970),
     139-148.   Also report CONF-700301

     The similarity in behaviour show by lithium-drifted germanium
and silicon detectors in recent detailed studies of their spectrometer performance over a wide temperature range suggested a
comparison with the effects of temperature on the operation of
other types of semiconductor detectors. A summary of the data
published in the literature on temperature effects in compensated (p-i-n) and non-compensated (p-n) detectors is presented.

504. Martini, M. and McMath, T. A. TRAPPING AND DETRAPPING EFFECTS IN
     LITHIUM-DRIFTED GERMANIUM AND SILICON DETECTORS. Nucl.
     Instrum. Methods, 79 (1970), 259-76.

     The trapping-detrapping behaviour of Ge(Li) and Si(Li) detec-
tors at low temperatures (T<77K) has been quantitatively analyzed
by means of the phenomenological theory published by J. W. Mayer.
The conclusion has been reached that the trapping centers are
the primary dopants: Li and Ga in Ge(Li) detectors, Li and B in
Si(Li) detectors.

505. Maslova, L. V.; Matveev, O. A.; Ryvkin, S. M.; Strokan, N. B.,
     and Khusainov, A. Kh. GAMMA SPECTROMETRY BASED ON GERMANIUM
     DETECTORS. Oak Ridge National Laboratory, Tenn. June 1966.
     6p.   ORNL-TR-1651; CONF-660625

     The preparation of electrical contacts on Ge(Li) detectors
is discussed. These contacts must be made very carefully in
order to obtain high resolution and minimum detector noise. It
is based on introduction of radiation defects in a germanium
crystal. The advantage of these detectors compared with Ge(Li)
detectors is the fact that they do not have to be stored at
liquid-nitrogen temperatures. The characteristics of these radiation detectors are preserved during storage at room temperature.

506. Maslova, L. V.; Matveev, O. A.; Ryvkin, S. M.; Strokan, N. B., and
     Sondaevskaya, I. A. GERMANIUM P-I-N DETECTOR WITH HIGH RE-
     SOLVING POWER FOR SMALL AND MEDIUM-ENERGY GAMMA QUANTA. At.
     Energ. (U.S.S.R.), 18 (June 1965), 654-5.

     Descriptions are given of the technology, construction and
     efficiency of germanium detectors using p-type germanium with
     specific resistance 3 to 10 ohm cm. The detector is designed
     for measuring gamma-quantum energies with amplitude resolving
     power 1.333 MeV. Diagrams are given of the design with the guard
     and connecting switch. The gamma spectra of ^{60}Co and ^{181}Hf and
     ^{137}Cs preparations are included.

507. Mathiesen, J. M. and Hurley, J. P. GERMANIUM (LITHIUM) DETECTOR
     FABRICATION. U. S. Naval Radiological Defense Laboratory,
     San Francisco, Calif. Dec. 9, 1965. 30p.
                         USNRDL-TR-949; Also report AD-627930

     A step-by-step description is given for each of the procedures
     in the fabrication of a Ge(Li) detector, including germanium pre-
     paration, lithium diffusion, lithium drift, and test and evalua-
     tion. The necessary materials and equipment are also listed to-
     gether with their availabilities. The test is augmented with de-
     tailed diagrams.

508. Matsuda, K.; Nonaka, I.; Omata, K.; Yagi, K., and Koike, M.
     ELECTRICAL DETECTION SYSTEM FOR BROAD-RANGE MAGNETIC SPEC-
     TROMETER. Nucl. Instrum. Methods, 53 (1967), 82-86.

     The performance of the electrical detection system used for
     the spectrum analysis of broad range magnetic analysers is de-
     scribed. Two kinds of detector array of 250 solid state detectors
     and the other is an array of 200 proportional-scintillation counter
     telescopes. They are followed with the same system of pulse
     handling. Informations are stored in an 1024 channel PHA core
     memory. Then the spectrum obtained can be processed with the PHA
     function.

509. Matveev, O. A. GERMANIUM GAMMA COUNTER. pp. 64-65 of "Inter-
     nationale Arbeitstagung Herstellung und Anwendung von Hal-
     bleiterdetektoren." Sept. 1963. (In Russian)   ZFK-PHA-12

     Germanium p-i-n junction detectors can be used as gamma spec-
     trometric counters with much higher efficiency than silicon detec-
     tors, because of the higher atomic number of Ge. Si detectors also
     require cooling, the Ge P-I-N junctions is described. The gamma
     spectrum of ^{137}CS measured using the GE detector is presented.

510. Mayer, J. W.; Zano, K. R.; Martini, M., and Fowler, I. L.
     INFLUENCE OF TRAPPING AND DETRAPPING EFFECTS IN SI(LI), GE
     (LI) AND CDTE DETECTORS. IEEE Trans. Nucl. Sci., NS-17,
     no. 3 (June 1970), 221-34.   Also report CONF-700301

     Carrier capture and re-emission effects in semiconductor
     nuclear particle detectors are characterized by a trapping time $\tau+$
     and detrapping time τD. Measurement of these quantities as a
     function of gibes information on capture cross-sections, σ, and
     activation energies, E_τ. Trapping and detrapping processes are
     field dependent. These effects have been investigated in Si(Li)
     and Ge(Li) detectors and are discussed in relation to operating
     temperature and detector resolution. The performance of Ge(Li)
     and CdTe detectors is evaluated in view of the influence of
     trapping effects.

511. Mayer, J. W. MATERIAL REQUIREMENTS FOR SEMICONDUCTOR DEVICES
     IN GAMMA RAY DETECTORS. pp. 1-10 of "Nucleonics in Aero-
     space." Polishuk, Paul, ed. New York, Plenum Press, 1968.
                                                  CONF-670714

     The requirements are presented for high Z semiconductor
     nuclear particle detector suitable for operation as a gamma de-
     tector at elevated temperatures.

512. Mayer, J. W. SEMICONDUCTOR DETECTORS FOR NUCLEAR SPECTROMETRY.
     Nucl. Instrum. Methods, 43 (1967), 55-64.

     A general description is presented of the "position" detec-
     tor and the avalanche detector including their uses and limita-
     tions. We discuss diverse factors concerned with detector per-
     formance as charge collection, energy resolution and Fano factor,
     energy required to form a hole-electron pair, fission fragment,
     pulse height defect, etc. Special attention is given to the
     choice of materials for use in gamma ray spectroscopy and the
     fabrication techniques involved.

513. Mayer, J. W. USE OF ION IMPLANTATION TECHNIQUES TO FABRICATE
     SEMICONDUCTOR NUCLEAR PARTICLE DETECTORS. Nucl. Instrum.
     Methods, 63 (1968), 141-51.

     A review of ion-implantation processes as applied to semi-
     conductor nuclear particle detectors. Discusses the role of
     channeling effects in determining the distribution of implanted
     atoms, defects and junction characteristics.

514. Megerth, F. H. ZIRCALOY-CLAD UO$_2$ FUEL ROD EVALUATION PROGRAM.
     General Electric Company, San Jose, California. Reactor
     Fuels and Reprocessing Dept., Quarterly Progress Report,
     no. 11. Aug. 1970. 52p.   GEAP-10217

     The relative rod-average burnups of 41 rods were determined
by gamma scanning with a germanium (lithium) detector system and
counting the ^{137}Cs line.

515. Meyer, H.; Eschbach, H. L.; Nagel, W., and Kruidhof, E. W. THE
     FABRICATION OF ENCAPSULATED GE(LI) DETECTORS OF GAMMA-SPECTRO-
     SCOPY. European Atomic Energy Community, Geel,(Belgium).
     Joint Nuclear Research Center. 1968. 22p.   EUR-4063e

     The procedure and the equipment for the fabrication of en-
capsulated Ge(Li) are discussed.

516. Meyer, H. and Verelst, H. PERFORMANCE TESTS FOR GE(LI) SPEC-
     TROMETERS. pp. 171-8 of "Ispra Nuclear Electronics Symposium.
     Proceedings." Stresna, Italy. May 1969.
                               EUR-4289; Also report CONF-690515

     Test equipment for the simulation of experiment conditions
was developed to simplify the design and to standardize the test
of high precision instrumentation for spectrometers in the future.
For linearity, stability and resolution tests an electronic precision pulse generator has been developed which allows an almost
perfect simulation of detector signals and measurement accuracy,
better than 0.01%. A random generator for the simulation of
Compton continua is described which has been applied together
with the precision pulser for high rate performance tests of a
spectrometer.

517. Meyer, H. and Verelst, H. SPECTRAL PERFORMANCE OF A GE(LI) PULSE
     HEIGHT SPECTROMETER AT HIGH EVENT RATES. European Atomic
     Energy Community, Geel, (Belgium). Joint Nuclear Research
     Center. Brussels. 1969. 12p.   EUR-4240e

     For gamma-ray spectroscopy with Ge(Li) detectors high event
rates are often desirable, not only for their application in
coincidence experiments but also because of the presence of large
Compton background. Modern instrumentation techniques were
applied and a fast servostabilized pulse height converter [14] has
been used for the measurement of gamma-ray spectra up to event
rates of 300 kHz. The features of the spectrometer are described
and performance limitations are discussed. Limitations are
introduced especially by the detectors, if high peak-rate-to-peak
width ratios are desired.

518. Meyer, O. and Haushahn, G. HIGH RESOLUTION PARTICLE-DETECTORS
     PRODUCED BY ION-IMPLANTATION. Nucl. Instrum. Methods, 56
     (1967), 177-178.

     High resolution particle detectors are produced by ion-im-
     plantation technique. It is shown that the improved performance
     is due to the fact that well-channeled ions at relative low
     energy and low ion-concentrations minimize bulk damage effects.

519. Meyer, O.; Baumgaertner, M., and Wuechner, F. PREPARATION OF
     CYLINDRICAL COAXIAL-DRIFTED GERMANIUM DETECTORS WITH LARGE
     ACTIVE VOLUME. Nukleonik, 10, no. 4 (1967), 189-92.

520. Meyer, O. SEMICONDUCTOR COUNTERS WITH THIN WINDOW N+ AND P+
     -CONTACTS PRODUCED BY ION-IMPLANTATION. IEEE Trans. Nucl.
     Sci., NS-15, no. 3 (June 1968), 232-238.

     Ion-implantation techniques were used to produce n+ and p+
     contacts on n and p-type germanium, n and p-type silicon and
     lithium compensated germanium. Interstitial and substitutional
     doping behaviour of 33 elements were investigated, only with
     few ions it is possible to get non-injecting contacts with good
     reverse characteristics using low temperature annealing process.
     Results on the application of those contacts to radiation de-
     tection are given.

521. Michaelis, W. EXPERIMENTAL GAMMA RAY RESPONSE CHARACTERISTICS OF
     LITHIUM-DRIFTED GERMANIUM DETECTORS. Nucl. Instrum. Methods,
     70 (May 1, 1969), 253-61.

     Experimental determinations of absolute intrinsic efficiency,
     absolute total absorption probability, and peak-to-total ratio
     of lithium-drifted germanium diodes for gamma rays between 0.1
     and 2 MeV are described for several detector and irradiation
     geometries. The sensitive volume of the crystals considered
     ranges from 3.8 to 28 cm^3. The results may be used as a guide
     in determining the optimum crystal dimensions and irradiation
     geometry for a given gamma-ray energy. The data are compared
     with simple models and, if available, with detailed theoretical
     calculations of the response characteristics.

522. Michaelis, W. and Kuepfer, H. HIGH-RESOLUTION GE(LI) ANTI-COMPTON
     SPECTROMETER FOR RADIATION NEUTRON CAPTURE SPECTROSCOPY.
     Nucl. Instrum. Methods, 56 (1967), 181-188.

A Ge(Li) anti-Compton spectrometer for the energy range up
to 3 MeV is described which allows for the special conditions
of (n, γ) spectroscopy and which gives improved performances
compared to previously reported devices. The system consists
of planar 4.9 cm^3, Ge(Li) diode, a 50 cm dia. x 40 cm plastic
scintillator and 4" dia. x 6" NaI(Tl) detector at scattering
angle 0°. The anti-coincidence method together with a pulse-
shape discrimination technique very effectively suppress the
Compton background under the peaks. The energy resolution is
2.15 keV fwhm at 0.662 MeV. Typical capture spectra are presented to illustrate the performance of the system.

523. Michaelis, W. and Schmidt, H. ON THE APPLICATION OF LITHIUM-
        DRIFTED GERMANIUM DIODES IN NEUTRON CAPTURE GAMMA RAY
        SPECTROSCOPY. Kernforschungszentrum, Karlsruhe, (W. Germany)
        Inst. fuer Angewandte Kernphysik. June 1966. 15p.
                                                    CONF-660625-2

A status report is presented on the application of lithium-
drifted germanium diodes in neutron capture gamma ray spectroscopy. The report includes a summary of the research being done
at Karlsruhe on neutron capture.

524. Michaelis, W. TABLES OF EXPERIMENTAL ABSOLUTE TOTAL ABSORPTION
        PROBABILITIES, INTRINSIC EFFICIENCIES AND PEAK-TO-TOTAL
        RATIOS FOR LITHIUM-DRIFTED GERMANIUM DIODES. Kernforschungs-
        zentrum Karlsruhe, (W. Germany) Inst. fuer Angewandte Kern-
        physik. Oct. 1968. 10p.   KFK-865

Experimental values for the absolute total absorption
probability, absolute intrinsic efficiency and peak-to-total ratic
of lithium-drifted germanium diodes for gamma rays between 0.1
and 2 MeV are presented for several detectors and irradiation
geometries. The sensitive volume of the crystals considered
ranges from 3.8 to 28 cm^3. Increasing attention is given to the
application of such devices in gamma-ray spectroscopy.

525. Miehe, J. A.; Siffert, P., and Coche, A. EFFECT OF DIFFERENT
        NOISE SOURCES ON THE TIME RESOLUTION OF GE(LI) DIODES. Nucl.
        Instrum. Methods, 75 (1969), 328-30. (In French)

Timing properties of planar and coaxial counters are studied.
The influence of both series and parallel noise sources is
clearly shown. Fwhm of the coincidence curves of 1.6 ns (^{22}Na)
for planar detectors and 3.4 ns for coaxial counters have been
obtained (w = 7 - 8 mm).

526. Miehe, J. A.; Siffert, P., and Coche, A. THE INFLUENCE OF
        DIFFERENT NOISE SOURCES ON TIME RESOLUTION OF GE(LI) DIODES.
        Nucl. Instrum. Methods, 75, no. 2 (Nov. 1969), 328-30.

527. Miehe, J. A.; Siffert, P.; Stuck, R., and Coche, A. INFLUENCE
        OF THE COMPENSATED THICKNESS OF COAXIAL GE(LI) DETECTORS
        AND OF NOISE SOURCES ON TIMING PROPERTIES. pp. 95-100 of
        "Ispra Nuclear Electronics Symposium. Proceedings." Stresa,
        Italy. May 1969.   Also report CONF-690515; EUR-4289

        Starting with equal diameter germanium crystals, double open
    ended Ge(Li) coaxial diodes were fabricated in which the compen-
    sated region is varied by 2 mm steps. The spread of the charge
    collection times and the time resolution are studied simultane-
    ously. The influence of other parameters, such as the trigger-
    ing level of the fast discriminator are analyzed. The results
    are compared to those of a planar counter having the same
    sensitive thickness. The influence of parallel and series
    noise sources are studied.

528. Miehe, J. A.; Ostertag, E., and Coche, A. ON THE USE OF GE(LI)
        DETECTORS IN SHORT TIME MEASUREMENTS. Onde Elec., 46 (July-
        Aug. 1966), 801-3. (In French)

        Preliminary results of resolving time measurements performed
    with an experimental arrangement using one Ge(Li) detector in
    coincidence with one scintillation counter are given. The prompt
    resolution curve shows a full width at half maximum of 3 ns.

529. Miehe, J. A. and Siffert, P. TIMING PERFORMANCE OF GE(LI) DE-
        TECTORS. IEEE Trans. Nucl. Sci., NS-17, no. 5 (Oct. 1970),
        8-12.

        The influence of several parameters related to the detector,
    such as depletion layer thickness, external radius of coaxial
    counters, applied electric field on the timing properties of
    planar and two open ended coaxial Ge(Li) detectors is investi-
    gated.

530. Miller, G. L.; Wagner, S., and Yuan, L. C. L. INVESTIGATIONS
        ON LITHIUM-DRIFT SOLID-STATE DETECTORS FOR HIGH-ENERGY
        PARTICLE DETECTION. Nucl. Instrum. Methods, 20 (Jan. 1963),
        303-5.

        Lithium-drift solid-state detectors with depletion layers of

several millimeters were developed. A lithium-drift detector,
1 cm in diameter and 2.3 mm thick, was placed in a momentum-
analyzed beam consisting of pions and protons with momenta of 3
and 4 Bev/c. Energy spectra at different bias voltages are
shown. Good resolution was obtained operating at room temperature.

531. Miller, G. L.; Pate, B. D., and Wagner, S. PRODUCTION OF THICK
     SEMICONDUCTOR RADIATION DETECTORS BY LITHIUM DRIFTING.
     IEEE Trans. Nucl. Sci., NS-11 (Jan. 1964), 220-229.

532. Miner, C. Eugene. MOUNTINGS AND HOUSINGS FOR LITHIUM-DRIFTED
     SILICON AND GERMANIUM DETECTORS. Lawrence Radiation Labora-
     tory, University of California, Berkeley. Feb. 24, 1965.
     41p.   UCRL-11946

     A versatile cooling and housing design and several methods
for mounting lithium-drifted silicon and germanium detectors
are discussed. Since considerable care must be exercised to
keep non-encapsulated germanium detectors in their best operating
condition, greater emphasis is placed on mountings for germanium
detectors than on mountings for silicon detectors. Developmental
problems, materials, and methods of evacuating the housings are
discussed in detail.

533. Miner, C. Eugene. A SEMICONDUCTOR DETECTOR CRYOSTAT. Nucl.
     Instrum. Methods, 55 (1967), 125-137.

     This report describes in detail the mechanical design of a
multipurpose cryostat system for non-encapsulated Ge(Li) and
Si(Li) semiconductor radiation detectors.

534. Monteith, Larry K. CORRELATION OF I-V CHARACTERISTIC WITH NOISE
     FOR ION DRIFTED P-I-N JUNCTION PARTICLE DETECTORS. Rev.
     Sci. Instr., 35 (Mar. 1964), 388-92.

     A technique is described for achieving minimum noise in a
p-i-n particle detector. The reverse saturation current of the
detector and its relative noise are correlated for different
surface treatments. The data show that the reverse current at
the detector operating voltage is not a sufficient measure of
relative noise. The relative noise can be predicted from the
reverse current at two to three times the operating voltage.

535. Mooney, J. B.   A REGROWN P+ CONTACT FOR LITHIUM-DRIFTED GERMANIUM
     GAMMA-RAY DETECTORS. Nucl. Instrum. Methods, 50 (1967),
     242-244.

     A method for applying regrown Ga-In contacts to lithium-
     drifted germanium gamma ray detectors, either before or after
     drifting, has been developed. Since alloying of the regrown
     contact and lithium diffusion can be accomplished during the same
     temperature cycle, it is possible to re-arrange contacts and
     thereby construct thin, deep, low-capacitance detectors from
     large-area shallow ones. These in turn can be combined in one
     package to make deep, large volume detectors.

536. Morgan, G.; Owen, G., and Lee, Y. K.   FABRICATION OF LARGE PLANAR
     GE(LI) DETECTORS. Nucl. Instrum. Methods, 76 (1969), 169-70.

     Several new techniques in the fabrication of large area,
     planar Ge(Li) detectors with thin windows are described, including
     the use of an argon atmosphere, a simplified drifting device, and
     gold sputtering for windows.

537. Morimitsu, Wataru and Ishizuka, Yasuhiro. MAKING OF GE(LI) DE-
     TECTOR AND ITS APPLICATION. Kanagawa-ken Kogyo Shikenjo
     Kenkyu Hokoku, no. 29 (Mar. 1970), 37-42. (In Japanese)

538. Moszynski, Marek.   ADAPTATION OF THE TIME-TO-AMPLITUDE CONVERTER
     TO TIME MEASUREMENTS WITH THE AID OF SEMICONDUCTOR RADIATION
     DETECTORS. p. 100 of "AEC Nukleonika, v. 13, no. 13." 1969.

     The adaptation of a time-to-amplitude converter for use with
     semiconductor detectors is described for measuring detector time
     spreads and coincidence in nuclear spectroscopy. The coincidence
     curves in the scintillation counter-Si(Li) or Ge(Li) detector are
     shown.

539. Moszynski, Marek and Bengtson, B. APPLICATION OF A PULSE SHAPE
     SELECTION METHOD TO A TRUE COAXIAL GE(LI) DETECTOR FOR MEAS-
     UREMENTS OF NANO-SECONDS HALF-LIVES. Nucl. Instrum. Methods,
     80 (1970), 233-8.

     A study of the pulse shape distribution from 35 cm^3 true co-
     axial Ge(Li) detector has been performed for uniform irradiation.
     Two well defined pulse-shape groups were found which could be
     separated completely by CR differentiation. The prompt time spec-
     trum derived from the earlier group of pulses gave a fast

and exponential slope over more than four decades. By this
method it was possible to identify a very weak and delayed
transition in the nano-second range.

540. Motz, H. T.   LITHIUM-DRIFTED GERMANIUM DETECTORS AT THE LOS
        ALAMOS SCIENTIFIC LABORATORY: STATUS REPORT. pp. 56-72 of
        "The Panel on Lithium-Drifted Germanium Detectors. Proceed-
        ings." Vienna, International Atomic Energy Agency, 1967.

      The fabrication of planar and coaxial lithium-drifted ger-
manium at Los Alamos Laboratory is described in detail.

541. MOUND LABORATORY CHEMISTRY AND PHYSICS PROGRESS REPORT, APRIL-
        JUNE 1970. Mound Laboratory, Miamisburg, Ohio. Nov. 20,
        1970. 21p.   MLM-1751

      Topics covered include ^{234}U recovery, ^{229}Th recovery, addi-
tion compound of UCl_4 with tetramethylurea.

542. Movchan, E. A.   PREPARATION OF CONTACTS ON N-TYPE GERMANIUM THAT
        ARE NEUTRAL IN STRONG ELECTRIC FIELDS. Instrum. and Experi-
        mental Tech., no. 6 (1967), 1457.

543. Mowatt, R. S.   A SEMI-EMPIRICAL EFFICIENCY CURVE FOR A GE(LI)
        DETECTOR IN THE ENERGY RANGE 50 TO 1400 KEV. Nucl. Instrum.
        Methods, 70 (May 1969), 237-44.

      A semi-empirical method is given to obtain the relative full-
energy-peak efficiency $\varepsilon\gamma$ of a Ge(Li) diode over the γ-ray energy
range 50 to 1400 keV. The method is an extension of the work
done by Freeman and Jenkin and accounts for both γ-ray attenuation in the materials present between a fixed source position
and the intrinsic region of the diode and the saturation of the
diode efficiency at low energies. The formula was fitted to the
observed relative efficiency of a 2 mm depleted Ge(Li) planar
diode to an accuracy of 1-1/2% over the range 100 to 1400 keV
and 4% over the range 50 to 1400 keV. The dependence of the
shape of the relative efficiency curve on source-to-diode distance was examined using a ^{169}Yb point source.

544. Mowatt, R. S. ^{152g}Eu and ^{226}Ra RELATIVE γ-RAY INTENSITIES FOR
        RAPID EFFICIENCY CALIBRATIONS OF GE(LI) DETECTORS. Can. J.
        Physics, 48 (Nov. 1, 1970), 2606-10.

      The relative intensities of γ-rays following the decay of

^{152}Eu and ^{226}Ra (in equilibrium with its decay products) were studied in order to obtain rapid efficiency calibrations of Ge(Li) detectors.

545. Muggleton, A H. F. LARGE VOLUME LITHIUM DRIFTED GERMANIUM DIODE FOR GAMMA RAY SPECTROSCOPY. Atomic Weapons Res., East Aldermaston, England. Oct. 1966. 20p.   AWRE-NR-3/66

Coaxial lithium drifted germanium diodes with sensitive volumes greater than 17 cc were manufactured and tested as γ-ray spectrometers. Detail fabrication techniques are described together with details of detector performance.

546. Muggleton, A. H. F. PRODUCTION OF LARGE VOLUME GE(LI) DETECTORS. pp. 121-31 of the "Proceedings of the Meeting on Special Techniques and Materials for Semiconductor Detectors, Ispra, Italy, 1968." June 1969.

Techniques and problems involved in manufacturing large Ge(Li) detectors on an industrial scale are described.

547. Mukherjee, P. and Sengupta, A. K. ^{152}EU AS CALIBRATION SOURCE FOR GE(LI) DETECTORS. Nucl. Instrum. Methods, 68 (Feb. 1, 1969), 165-6.

The relative intensities of ten prominent gamma rays of ^{152}Eu (12.4y) are given which can be used to obtain promptly the relative photopeak efficiency of a Ge(Li) gamma detector.

548. Mullady, James Benedict. FABRICATION OF LITHIUM-DRIFT GERMANIUM RADIATION DETECTORS. Air Force Inst. Tech., Wright-Paterson AFB, Ohio. School of Engineering. June 1966. 55p. Thesis
                                                Also report AD-65248

A simple technique for fabricating lithium-drift germanium radiation detectors for use in gamma-ray spectroscopy is described in detail.

549. Nadkarni, R. A. and Ehmann, W. D. INSTRUMENTAL NEUTRON ACTIVA- TION ANALYSIS OF TOBACCO PRODUCTS. pp. 190-196 of "Modern Trends in Activation Analysis." vol. 1. Washington, D. C., National Bureau of Standards, 1969.

In this study instrumental neutron activation analysis (INAA)

utilizing a high resolution Ge(Li) detector has been employed.
This detector permits resolution and characterization of close
lying gamma-ray energies. The work presented here is confined
to the abundances of 11 elements in tobacco. The studies on the
transferrence of these elements into smoke and residual ashes
will be reported at a later date.

550. Nagahara, Teruaki and Ishizuka, Yasuhiro. CONSTRUCTION OF
     A GERMANIUM P-I-N DETECTOR. Bunseki Kiki, 9 (Nov. 1968),
     707-15. (In Japanese)

     Procedures for fabricating a germanium p-i-n detector are
described. The selection and processing of germanium and the
lithium-drifting process are discussed. Voltage and current
characteristics, capacity of the detector, energy resolution,
and detection efficiency are also discussed.

551. Nagpal, T. S. and Gaucher, R. E. ^{75}Se AS A CALIBRATION STANDARD
     FOR GE(LI) DETECTORS. Nucl. Instrum. Methods, 89 (1970),
     311-13.

     The use of ^{75}Se for calibrating Ge(Li) detectors is dis-
cussed. An example is given using its accurately measured energies and intensities.

552. Neeb, K. H.; Franke, R., and Neidl, H. USE OF A GE(LI) DETECTOR
     IN RADIOCHEMICAL ANALYSIS. 1. ACTIVATION ANALYSIS OF PPM A-
     MOUNTS OF TELLURIUM IN SELENIUM. Z. Anal. Chem., 247 (Oct.
     30, 1969), 225-8. (In German)

     The very high energy resolution of Ge(Li) detectors in γ
spectrometry enables chemical separation processes to be shorten-
ed quite considerably in many cases of radiochemical analysis.

553. Negrei, S. A. and Firsov, E. I. A GE(LI)-SPECTROMETER FOR THE
     STUDY OF THE GAMMA RADIATION DUE TO THE RADIATIVE CAPTURE
     OF THERMAL NEUTRONS. Vestsi. Akad. Navuk BSSR, Ser. Fiz-Mat.
     Navuk, no. 2 (1970), 128-31. (In Russian)

     A gamma spectrometer provided with a Ge(Li) detector posses-
sing a sensitive volume of 2.7 cm^3 is described. The instrument
was installed in the tangential beamhole of the IRT-2000 reactor
of the Belorussian Academy of Sciences for the purpose of studying the gamma spectra resulting from (n, γ) reactions. The
lithium was deposited on the surface of the germanium crystal by

sputtering and diffused into it at 420°C. The device had a
resolution of 5.75 and 6.3 MeV for the 1.173 and 1.333 MeV lines
of a ^{60}Co source, respectively. It was used for studying the
gamma-ray spectrum of ^{36}Cl from the ^{35}Cl (n, γ) ^{36}Cl reaction.

554. Negrei, S. A. and Firsov, E. L. SEMICONDUCTOR GAMMA SPECTROMETER.
    Instrum. Exp. Tech. (USSR), no. 1 (English Transl.) (Jan.-
    Feb. 1970), 45-7.

    A gamma spectrometer is described having a Ge(Li) detector
2.7 cm^3 in volume with a resolution of 0.5% for ^{60}Co gamma radiation. The technique for preparing planar detectors is described.

555. Nelson, J. B.; Blake, K. R.; Mitchell, J. C.; Allen, S. J., and
    Hunt, W. D. FAST/THERMAL NEUTRON RATIOS DETERMINED BY GE(LI)
    GAMMA SPECTROMETRY. pp. 334-40 of the "International Sym-
    posium on Nucleonics in Aerospace." Columbus, Ohio. 1967.
                                                    CONF-670714

    Fast to thermal neutron ratios are often determined by meas-
urement of the cadmium in/out gamma ratio using the ^{197}Au (n, γ)
reaction. However, the ^{197}Au(n, ^{2}n) reaction competes strongly
in the presence of 14 MeV neutrons and limits the usefulness of
gold foil when counted in a NaI gamma-ray spectrometer. A method utilizing a high resolution Ge(Li) spectrometer whereby the
fast and thermal neutron fluence can be measured using gold
foil only is described.

556. Nitsche, E. UNIVERSAL VACUUM CHAMBER FOR LITHIUM-DRIFTED GERMAN-
    IUM DETECTORS. pp. 137-142 of "The Proceeding of the Meeting
    on Special Techniques and Materials for Semiconductor Detec-
    tors, Ispra, Italy, 1968." Capellani, F., ed. June 1969.
                            EUR-4269; Also report CONF-681049

    The design of a vacuum chamber of small outer dimension for
encapsulating germanium detectors of different shapes and sizes
to be mounted horizontally or vertically is described. The vacuum chamber forms a part of a cryostat system and in certain applications it is advantageous to use a germanium detector of the
suspended type.

557. Notea, A. THE GE(LI) SPECTROMETER AS A POINT DETECTOR. Nucl.
    Instrum. Methods, 91 (1971), 513-515.

    The behaviour of a Ge(Li) spectrometer as a point detector was

considered. The intrinsic peak efficiency of the point detector
is suggested as a characteristic parameter. A procedure for
relative calibration of photon sources, and for comparison between various shape spectrometers, based on accurate corrections
of solid angle as a function of photon energy is presented.
Examples of the procedure are given.

558. Novikov, S. R.,; Novogrudskii, B. V., and Pustovoit, A. K.
      COLLECTION OF CHARGE IN γ-RAY COUNTERS MADE OF LITHIUM-
      DRIFTED GERMANIUM. Sov. Phys.-Semicond. (Eng. Transl.) 4
      (Dec. 1970), 887-91.

      The relationship between the effective value of the carrier
      lifetime (τ_{eff}) found from the charge collection in lithium-
      drifted germanium counters and the values of the lifetime deter-
      mined by various methods in the initial material is analyzed.
      For most of the investigated crystals, the value of τ_{eff} is
      correlated with the 78°K value of the initial lifetime, which
      was either deduced from the short-circuit current of a diode
      prepared from the initial material and irradiated with γ-ray
      quanta, or measured by the injection-extraction method.

559. Nurmia, Matti and Stubb, Tor. PREPARATION AND USE OF GE(LI)
      GAMMA DETECTORS. Ann. Acad. Sci. Fenn. Ser. A, VI, no. 213
      (1966), 1-10.

      Techniques used in the preparation of lithium-drifted ger-
      manium detectors are described. The lithium drifting process is
      controlled with a thermoelectric heating and cooling module.
      The method has yielded detectors with a thickness of up to 2.5 mm
      and a resolution limited by the preamplifier noise (under 5 keV
      FWHM).

560. O'Hanlan, J. F. and Hoering, R. R. CaF_2 COMPENSATION OF SURFACE
      STATES IN GE(LI) DETECTORS. Nucl. Instrum. Methods, 53
      (1967), 341-345.

      A CaF_2 vacuum deposition technique is described for setting
      surface states in Ge(Li) detectors. Evidence is discussed for
      the protection of the junction edges from the effects of exposure
      to atmospheric ambients.

561. Okubo, Dan. DOUBLE-DRIFT THIN-WINDOW PLANAR GE(LI) DETECTORS.
      University of California, Livermore, Lawrence Radiation
      Laboratory. Oct. 16, 1969. 8p. UCRL-71787; CONF-691017-2

From the 16h Nuclear Science Symposium, San Francisco, California.

An improved method for fabricating a thin-window planar detector was developed at Lawrence Radiation Laboratory, Livermore. This detector is designed to analyze spectra of low-energy as well as high-energy gamma and x-rays. The detector has planar configuration with a relatively large gold surface barrier entry window. The main advantage of this detector is its faster drifting time coupled with minimum waste of material.

562. Oria, M.; Ripon, R., and Lepetit, J.   CONSTRUCTION AND FOCUSING OF A GE(LI) COMPTON SPECTROMETER. RC Accad. Naz. Lincei, Italy, 45, no. 5 (Nov. 1968), 267-8. (In French)

A single crystal Ge(Li) Compton spectrometer is described and has been tested.

563. Oria, M.; Ripon, R., and Lepetit, J.   DEVELOPMENT AND ADJUSTMENT OF A GE(LI) COMPTON SPECTROMETER. Rev. Phys. Appl., 4 (June 1969), 267-8. (In French)   Also report CONF-681208

From the Conference on Experimental Methods in Nuclear and Particle Physics, Strasbourg, France.

A single crystal Ge(Li) Compton spectrometer is described and has been tested.

564. Orphan, V. J. and Rasmussen, N. C.   A GE(LI) SPECTROMETER FOR STUDYING NEUTRON CAPTURE GAMMA RAYS. Nucl. Instrum. Methods, 48 (1967), 282-295.

A gamma-ray spectrometer, using a 30 cm^3 coaxial Ge(Li) detector, which can be operated as a pair spectrometer at high energies and in the Compton suppression mode at low energies provides an effective means of obtaining thermal neutron capture gamma spectra over nearly the entire capture gamma energy range. The energy resolution (fwhm) of the spectrometer is approximately 0.5% at 1 MeV and 0.1% at 7 MeV.

565. Orphan, V. J. and Rasmussen, N. C.   A PAIR SPECTROMETER USING A LARGE COAXIAL LITHIUM-DRIFTED GERMANIUM DETECTOR. IEEE Trans. Nucl. Sci., NS-14 (Feb. 1967), 544-558.

566. Ostertag, E.; Miehe, J. A.; Henck, R., and Siffert, P.   COINCIDENCE EXPERIMENTS WITH COAXIAL GE(LI) DETECTORS IN THE LOW

ENERGY RANGE. <u>IEEE Trans. Nucl. Sci.</u>, NS-15, no. 3 (June 1968), 413-418.

The γ-γ coincidence arrangement employing two large volume (20 cm^3) Ge(Li) coaxial detectors has been realized. Each diode has two isolated outputs, one of which yields the energy information of the incident radiation and the other one the timing signal.

567. Ostertag, E.; Miehe, J. A., and Coche, A. OBTENTION PROCESSES OF TEMPORAL INFORMATION IN A GE(LI) DIODE: APPLICATION TO GAMMA SPECTROMETRY. <u>Onde Elec.</u>, 48 (June 1968), 583-7. (In French)

The time at which a gamma particle is detected by a Ge(Li) semiconductor counter can be derived either from its current pulse or its integrated charge pulse. A Ge(Li)-Ge(Li) coincidence arrangement is described. The coincidence gamma-ray spectrum of ^{60}Co obtained in one detector, when the other is gated on the 1.333 keV full energy peak and when the coincidence resolving time is equal to 30 ns, has a threshold of 150 keV.

568. Owen, G. E. and Lee, Y. K. GAMMA RAY POLARIMETERS WITH GE(LI) DETECTORS. <u>Nucl. Instrum. Methods</u>, 82 (1970), 173-7.

A planar Ge(Li) detector is used in three different configurations as polarimeter. A thin planar detector by itself is found to be sensitive to linearly polarized gamma rays. A single detector partially sliced into three independent regions may be used as a polarimeter in two different ways. In all three cases the good resolution of Ge(Li) detectors is fully utilized.

569. Pagden, I. M. H.; Pearson, G. J., and Beck, V. N. SEMI-AUTOMATED COMPUTER SYSTEM FOR GAMMA RAY SPECTRUM ANALYSIS OF THERMAL NEUTRON-ACTIVATED SAMPLES. <u>IEEE Trans. Nucl. Sci.</u>, NS-17, no. 1 (Feb. 1970), 211-17. Also report CONF-691017-PT. 1

A set of consecutively executed computer programmes has been developed for the quantitative interpretation of gamma ray spectra obtained with Ge(Li) detectors. The method compares spectral information with catalogued nuclear data and yields least squares fitted sets of activity which subsequently may be converted to parent masses.

570. Palms, J. M.; Venugopala Rao, P., and Wood, R. E. THE CHARAC-
     TERISTICS OF AN ULTRA-HIGH RESOLUTION GE(LI) SPECTROMETER
     FOR SINGLES AND COINCIDENCE X-RAY AND GAMMA-RAY STUDIES.
     IEEE Trans. Nucl. Sci., NS-16, no. 1 (Feb. 1969), 36-46.

     The use of an ultra-high resolution Ge(Li) photon spec-
     trometer in low and high energy x-ray and gamma-ray measurements
     has been evaluated. Studies were made of the resolution, low
     energy cutoff, the full energy and K X-ray escape efficiency,
     the effective detector dead layer and the detector linearity.
     The detector was used simultaneously with a high resolution
     Si(Li) detector to measure fast coincidence between Ka_1, Ka_2,
     and LX-rays, as well as coincidences between K. L conversion
     electrons and LX-rays.

571. Palms, J. M.; Wood, R. E., and Puckett, O. H. A GE(LI) CONCEN-
     TRIC DUODE+ SPECTROMETER FOR COMPTON SUPPRESSION. IEEE
     Trans. Nucl. Sci., NS-15, no. 3 (June 1968), 397-406.

572. Palms, J. M. and Greenwood, Arthur. THERMOELECTRIC CONTROL
     APPARATUS FOR THE FABRICATION OF THICK LITHIUM-DRIFTED GER-
     MANIUM DETECTOR. Rev. Sci. Instrum., 36 (Aug. 1965), 1209-
     1213.

     Lithium-drifted germanium radiation detectors were fabricated
     using thermoelectric cooler-heater modules to provide temperature
     stabilization during the drifting period.

573. Palms, J. M.; Venugopalo Roa, P., and Wood, R. E. AN ULTRA-HIGH
     RESOLUTION GE(LI) SPECTROMETER FOR SINGLES AND COINCIDENCE
     X-RAY AND GAMMA-RAY STUDIES. Nucl. Instrum. Methods, 64
     (1968), 310-316.

     This paper describes the characteristics of a small (8 mm dia.,
     depletion depth 4 mm) ultra high resolution (FWHM-450eV at 14 keV
     and 1.7 keV at 1332 keV) Ge(Li) spectrometer. In particular the
     following were studied: resolution versus energy, the Fano Fac-
     tor for Ge, the full energy and KX-ray escape efficiency and the
     detector linearity.

574. PANEL ON THE USE OF LITHIUM-DRIFTED GERMANIUM GAMMA-RAY DETECTORS
     FOR RESEARCH IN NUCLEAR PHYSICS, VIENNA, 1966. Proceedings.
     Vienna, International Atomic Energy Agency, 1966.

The proceedings of a panel held in Vienna, 1966. Includes
papers on the fabrication and performances of the lithium-
drifted germanium detectors. Each paper is in its original
language with an abstract in English.

575. Paradellis, T. and Hontzeas, S. SEMI-EMPIRICAL EFFICIENCY EQUA-
     TION FOR GE(LI) DETECTORS. Nucl. Instrum. Methods, 73
     (1969), 210-14.

     A semi-empirical equation to describe the relative photopeak
     efficiency of Ge(Li) detectors (of various sizes and shapes) in
     the gamma-ray energy range of 200 to 1500 keV is presented.

576. Paris, P. and Treherne, J. PROTECTION DES BORDS DE DETECTEURS
     AU GERMANIUM. Nucl. Instrum. Methods, 63 (1968), 123-124.

     The compensated region of a Ge(Li) detector has been coated
     with a thin plastic film. The detector, cooled to 77°K, exhib-
     its good energy resolution. Such a protection decreases the
     effects of superficial stains.

577. Parker, C. V., Jr.; Martin, T. C., and Morgan, I. L. SEPARATION
     OF ISOTOPIC INTERFERENCES UTILIZING GE(LI) GAMMA RAY DETEC-
     TORS. Texas Nuclear Corp., Austin, Texas. July 1966. 4p.
                                                    ORO-2980-13

     A method that has proven successful in separating the partic-
     ularly troublesome interference between the 1.78 MeV gamma ray
     from ^{28}Si (n, P) ^{28}Al 1.81 MeV gamma ray from ^{56}Fe(n, p) ^{56}mn is
     presented. A lithium-drifted germanium solid state detector is
     used as a high resolution gamma spectrometer to provide adequate
     separation between these two gamma rays, after irradiation with
     14 MeV neutrons, the samples were counted using a 4.4 cc and
     13 cc Ge(Li) detector.

578. Parker, Jack L.; Holm, Dale M., and Barnes, Barry K. CHARACTERIS-
     TICS AND APPLICATIONS OF A LARGE SODIUM IODIDE DETECTOR ASSEM-
     BLY. pp. 1075-80 of "Modern Trends in Activation Analysis."
     Vol. 2. Washington, D. C., National Bureau of Standards, 1969.

     Response functions have been measured for a large sodium
     iodide double-crystal detector assembly. The assembly, used by
     itself and in conjunction with lithium-drifted germanium detec-
     tors, has been studied for applications in activation analysis
     and gamma-ray spectroscopy.

579. Parr, R. M. THEORETICAL PHOTOPEAK COUNT-RATES IN GE(LI) AND
     NAI(TL) DETECTORS FROM GAMMA-RAY EMITTERS FORMED BY THERMAL
     NEUTRON ACTIVATION. pp. 150-63 of "Uses of Activation Analy-
     sis in Studies of Mineral Element Metabolism in Man." Vienna,
     International Atomic Energy Agency, 1970.
                                          IAEA-122; CONF-680668

580. Pascal, Andre. METHOD OF MEASUREMENT FOR EVALUATING THE TRANS-
     PARENCY OF ENTRY WINDOWS OF GE(LI) SEMICONDUCTOR TO γ AND
     X PHOTONS. Commissariat a l'Energie Atomique, Saclay,
     France. Centre d'Etudes Nucleaires. July 1970. 16p. (In
     French)   CEA-N-1320

581. Passerieux, J. P. and Fouan, J. P. ASSEMBLY FOR DETECTION OF
     CHARGED PARTICLES AND GAMMA RAYS. Commissariat a l'Energie
     Atomique, Saclay (France). n.d. 3p. (In French)
                                               CEA-CONF-1155

     From the International Symposium on Nuclear Electronics,
     Versailles, France.
     The assembly for the detection of charged particles uses
     semiconducted detectors placed in the scattering chamber. The
     detectors, the preamplifiers, and the analysis amplifier are
     briefly described. For the detection of gamma rays the assembly
     uses either Ge(Li) coaxial diodes or NaI(Tl) scintillators.

582. Paugger, P. EXPERIENCE AND PROBLEMS IN THE INDUSTRIAL PRODUCTION
     OF GE(LI) DETECTORS. pp. 133-36 of the "Proceedings of the
     Meeting on Special Techniques, and Materials for Semiconductor
     Detectors, Ispra, Italy, 1968." June 1969.   EUR-4269

     The production of surface barrier and lithium-drifted ger-
     manium detectors at Nucletron is discussed. The system necessary
     for spectrometry has been developed and the FET preamplifier has
     a resolution of 1 to 2 keV for zero input capacitance and a
     slope of 0.02 keV/pF.

583. Pehl, R. H.; Goulding, F. S.; Landis, D. A., and Lenzlinger, M.
     ACCURATE DETERMINATION OF THE IONIZATION ENERGY IN SEMICON-
     DUCTOR DETECTORS. Nucl. Instrum. Methods, 59 (1968), 46-55.
                                          Also report CONF-670520

     Precise measurements were performed in order to determine the
     ionization energy for electrons and alpha particles in Ge and Si
     detectors. The data for Ge detectors indicate the ionization

energies for both particles are the same. Possible reasons for the difference found in Si detectors are suggested.

584. Pehl, R. H.; Landis, D. A., and Goulding, F. S. DETECTION OF HIGH ENERGY PROTONS BY THIN-WINDOW LITHIUM-DRIFTED GERMANIUM COUNTERS. IEEE Trans. Nucl. Sci., NS-13, no. 3 (June 1966), 274-277.   Also report UCRL-16722

Recent developments in producing thin-window Li-drifted germanium counters have enabled us to investigate their response to 29- and 40- MeV protons. The window is of negligible thickness for long-range particles about 0.5μ, although a precise measurement has not been made. Resolutions of 28 and 44 keV (FWHM) were obtained on 29- and 40-MeV protons respectively.

585. Pehl, R. H.; Landis, D. A.; Goulding, F. S., and Jarrett, B. V. DETECTION OF 30 AND 40 MEV PROTONS BY A THIN-WINDOW LITHIUM-DRIFTED GERMANIUM COUNTER. Phys. Lett., 19 (Dec. 1, 1965), 495-6.

The response of thin-window lithium-drifted germanium counters is investigated. Corrections were made for the spread of the beam energy, the thin window, and electronic noise. The remaining spread was found to be 21 keV for 30-MeV protons, and 30 keV for 40-MeV protons. Also the theoretical limitation of the energy resolution is given as a function of the energy deposited in a germanium crystal for different non-detector contributions.

586. Pehl, R. H. and Goulding, F. S. RECENT OBSERVATIONS OF THE FANO FACTOR IN GERMANIUM. Nucl. Instrum. Methods, 81 (1970), 329-30.

Results demonstrating that the Fano factor for germanium is less than 0.08 are presented.

587. Pehl, R. H.; Goulding, F. S.; Hansen, W. L., and Cordi, R. C. THIN-WINDOW GERMANIUM DETECTORS: FALLACY AND FACT. Nucl. Instrum. Methods, 75 (1969), 175-7.

Results demonstrating the successful use of a gold surface-barrier as a thin-window p+ termination of the intrinsic region of a lithium-drifted germanium detector are reviewed. These results are a rebuttal to statements recently made by Dearnaley and others in this journal.

588. Pepelnik, R.; Meyer, U., and Hick, H.   SYNCHRONIZED COMPTON
     SEMICONDUCTOR SPECTROMETER (SUMMIERENDES HALEITER-COMPTON-
     SPEKTROMETER).   Oesterreichische Studiengesellschaft fuer
     Atomenergie G. m. b.H., Seibersdorf, Austria. 1967. 13p.
     (In German)   SGAE-PH-55/1967

     The conventional Compton spectrometer measures electron
     energy with a semiconducting detector; motions of bound elec-
     trons cause spectral line broadening. The energy dependence
     of this effect is measured for germanium and silicium from 200
     keV to 2.5 MeV. Measurements with a summation Compton spectrom-
     eter that uses two Ge(Li)-detectors, eliminated this broadening
     effect.

589. Perkons, A. K. and Jervis, R. E.   APPLICATIONS OF HIGH-RESOLUTION
     GE(LI) GAMMA SPECTROMETRY IN CRIMINALISTICS.   Trans. Amer.
     Nucl. Sci., 11 (June 1968), 82-3.

590. Pervashin, Ju A. and Tolpygo, K. B.   PHOTOCONDUCTIVITY OF DE-
     FORMED SEMICONDUCTORS OF GE AND SI TYPE.   Soviet Physics-
     Semiconductor, 2, no. 3 (Sept. 1968), 324.

591. Petri, H. and Riepe, G.   SEMICONDUCTOR SPECTRA OF URANIUM FISSION
     PRODUCT MIXTURES. (HALBLEINTESPEKTREN VON URAN-SPALTPRODUKT-
     GEMISCHEN). pp. 164-170 of "Soc. Europeanne de Protec. and
     Radiations Measurement Tech. Sept. 1967." 1967. (In Ger-
     man)

     For routine determination of fission products (burn-up
     measurements) or fission gases (released in fuel elements) with
     scintillation counters, the spectrometry with Ge(Li) detectors
     as preorientation was tested.

592. Petushkov, A. A.   POSSIBILITIES OF THE USE OF SEMICONDUCTOR DE-
     TECTORS OF IONIZING RADIATION IN RADIOBIOLOGY AND MEDICINE.
     Radiobiologiya, 8 (1968), 135-41. (In Russian)

     A survey of the literature (31 references, 1949 to 1966) on
     applications of semiconductor detectors of ionizing radiation in
     radiobiology and medicine is presented. The most suitable were
     found to be those based on silicon and germanium with p-n and
     p-i-n junctions (small dimensions, high energy resolution, low
     pulse power consumption).

593. Peyrard, Michelle. REALISATION ET UTILISATION D'UN DETECTEUR
     GAMMA GERMANIUM POUR ETUDES SPECTROSCOPIQUES DE LA CHAINE
     DE DESINTEGRATION ^{171}HF→^{171}LU→^{171}YB. (DEVELOPMENT AND USE
     OF A GERMANIUM GAMMA DETECTOR FOR SPECTROSCOPIC STUDIES OF
     THE DECAY CHAIN ^{171}HF→^{171}LU→^{171}YB.) Grenoble Univ., France.
     Faculte des Sciences. 1965. 54p. Thesis. (In French)
                                                    Also report NP-17220

     The general properties of semiconductors are briefly out-
     lined. The techniques used for the preparation of a Li-compen-
     sated Ge detector are described in detail.

594. Phelps, Paul L. GAMMA-RAY SPECTROMETERS FOR THE ASSAY OF COM-
     PLEX MIXTURES OF LOW CONCENTRATION OF RADIONUCLIDES IN EN-
     VIRONMENT AND BIOLOGICAL MATERIALS. IEEE Trans. Nucl. Sci.,
     NS-15, no. 1 (Feb. 1968), 376-382.

595. Phelps, Paul L.; Hamby, Keith O.; Shore, B., and Potter, G. D.
     GE(LI) GAMMA-RAY SPECTROMETERS OF HIGH SENSITIVITY AND
     RESOLUTION FOR BIOLOGICAL AND ENVIRONMENTAL COUNTING. Uni-
     versity of California, Livermore, Lawrence Radiation Lab.
     May 24, 1968. 39p.   UCRL-50437

596. P-I-N DIODES PINPOINT GAMMA RADIATION. Can. Chem. Process, 49
     (Feb. 1965), 77-82.

     A specially-treated semiconductor promises to provide the key
     to better neutron activation analysis. Lithium-drifted germanium
     p-i-n diodes developed for gamma spectroscopy allow intrinsic
     efficiencies of 0.1 to 1.0% with resolutions that can be 10
     times better than the sodium iodide spectrometer.

597. Picard, Jean and Souchere, Gerard, PERTE D'ENERGIE, PARCOURSE ET
     RATENTISSEMENT DES PARTICULES CHARGES DANS LES DETECTEURS AU
     SILICIUM OU AU GERMANIUM. (ENERGY LOSS RANGE, AND STOPPING
     POWER OF CHARGED PARTICLES IN SILICON OR GERMANIUM DETECTORS.)
     Commissariat a l'Energie Atomique, Saclay, (France). Oct.
     1967. 25p. (In French)

     Energy loss range and stopping power of charged particles in
     silicon and germanium solid state detectors are presented in
     graph and table forms.

598. Pigneret, J.; Samueli, J. J., and Sarazin, A. THEORETICAL AND
     EXPERIMENTAL RESULTS ON THE RESOLVING TIME ON P-I-N DETECTORS

IN γ-γ COINCIDENCE EXPERIMENTS. IEEE Trans. Nucl. Sci.,
NS-13, no. 3 (June 1966), 306-314.

A calculation of the resolving time one can obtain by the
delayed coincidence method with a p-i-n detector is developed.

599. Platt, Robert B. EFFECTS OF RADIATION ON PLANT AND ANIMAL COM-
     MUNITIES. Progress Report, Jan. 1970 - Jan. 1971. Emory
     University, Atlanta, Ga. 1971. 102p.   ORO-2412-35

     This study discusses the design and performance of a NaI
anticoincidence-shielded Ge(Li) spectrometer for quantitative
measurements of low-activity radioisotopes in environmental
samples.

600. Poenaru, D. N.; Stuck, R., and Siffert, P. COLLECTION EFFICIENCY
     AND CHARGE CARRIER LOSSES IN COAXIAL AND PLANAR GE(LI) DETEC-
     TORS. INFLUENCE ON THE TIME RESOLUTION. IEEE Trans. Nucl.
     Sci., NS-17, no. 3 (June 1970), 176-186. Also rept. CONF-700301

     A theoretical study of the charge collection properties of
coaxial Ge(Li) detectors is presented.

601. Poenaru, D. N. COLLECTION TIME OF ELECTRON-HOLE PAIRS IN A
     COAXIAL GE(LI) RADIATION DETECTOR. IEEE Trans. Nucl. Sci.,
     NS-14, no. 1 (Oct. 1967), 1-7.

     Current and electric charge pulse shapes, obtained for the
collection of only one electron-hole pair released by an ionizing
event in the sensitive volume of coaxial Ge(Li) detector cooled
at liquid nitrogen temperature were calculated.

602. Poenaru, D. N. CURRENT AND CHARGE PULSES GIVEN BY COAXIAL GE(LI)
     RADIATION DETECTORS. Rev. Roum. Phys., 12 (1967), 951-62.

603. Poenaru, D. N. CURRENT AND VOLTAGE PULSES GIVEN BY SEMI-CONDUC-
     TOR RADIATION DETECTOR. Nucl. Instrum. Methods, 54 (1967),
     229-241.

     Voltage and current waveforms given by different types of
semi-conductor detectors on various load impedances were calculated.

604. Poenaru, D. N. and Vilcov, N. RADIATION DETECTORS WITH SEMI-
     CONDUCTOR DEVICES. PART 2. CHARACTERISTICS OF SEMICON-
     DUCTOR DETECTORS. Acad. Repub. Pop. Rom. Stud. Cercet.
     Fiz., 17 (1965), 695-723. (In Rumanian)

605. Poenaru, D. N. SEMICONDUCTOR RADIATION DETECTORS USED WITH
     CURRENT-SENSITIVE PREAMPLIFIERS. Nucl. Instrum. Methods,
     54 (1967), 242-249.

     The utilization of short current pulses given by L- and U-
     type semiconductor radiation detectors was investigated. The
     maximum amplitude and minimum length of the current pulses from
     silicon detectors were calculated and plotted against resistivi-
     ty for L detectors and voltage for U detectors respectively.
     By using a low input impedance, fast transistor current-
     sensitive preamplifier, L-detector exponential current wave-
     forms (for 4 MeV protons, 5-15 MeV alpha-particles and fission
     fragments) were verified. The possibility to employ current
     pulse for heavy charged particle spectrometry and to obtain
     simultaneously fast and slow pulses from the same detector were
     also checked experimentally.

606. Pohl, P. PROTECTIVE ALARM FOR LITHIUM DRIFTED GERMANIUM DETEC-
     TORS. Rev. Sci. Instrum., 41 (Aug. 1970), 1253-4.

     An alarm circuit for warning against excess leakage current
     in a lithium-drifted semiconductor detector due to inherent
     detector failure, vacuum failure, or cooling failure is des-
     scribed.

607. Polish Academy of Science, Warsaw. Institute of Nuclear Re-
     search. PREPARATION OF GERMANIUM GAMMA-SPECTROMETER BY
     LITHIUM DRIFT METHOD. (Polucheniye Germaniyevykh-spektrome-
     trov Gamma Izlucheniya po metodu dreifa ionov litiya). 1964.
     14p.   Rept.-592-11/PL

     Germanium diodes for gamma ray spectrometry using the lithium
     ion drift technique and p-type germanium have been fabricated
     active depths up to 1.5 mm were obtained.

608. Potter, Gilbert D.; McIntyre, David R., and Pomeroy, Deborah.
     TRANSPORT OF FALLOUT RADIONUCLIDES IN THE GRASS-TO-MILK FOOD
     CHAIN STUDIED WITH GERMANIUM LITHIUM-DRIFTED DETECTOR.
     Health Phys., 16 (Mar. 1969), 297-300.

A study demonstrating the use of a germanium lithium drifted detector in the field of radioecology.

609. Prasad, K. G.; Sharma, R. P., and Thosar, B. V. LEVEL STRUCTURE OF ^{199}AU. pp. 204-208 of "India. Dept. of Atomic Energy Proc. of the Nucl. Phys. and Solid State Physic Symposium Nucl. Physic." 1966.

A study of energy levels of ^{199}Au with the help of a Ge(Li) and a surface barrier detector. Includes a graph showing the gamma spectrum on the Ge(Li) detector.

610. Pratt, B. and Friedman, F. DIFFUSION OF LITHIUM INTO GE AND SI. J. Appl. Phys., 37, no. 4 (Mar. 15, 1966), 1893-96.

611. Pratt, T. A. E. C. MUONIC X-RAY EFFICIENCY CALIBRATION OF A GER-MANIUM-LITHIUM DRIFTED DETECTOR. Nucl. Instrum. Methods, 66 (Dec. 15, 1968), 348-350.

Efficiency calibration of a Ge(Li) detector using muonic x-rays from targets of various proton content is being discussed. After appropriate corrections a least-squares fit was made to the energy region 296 keV - 3435 keV. The ratio of Kα muonic X-ray to all muonic K X-rays is also reported.

612. Pratt, T. A. E. C. and Luther, M. L. STUDY OF GERMANIUM-LITHIUM DRIFTED GAMMA SPECTRUM SHAPES. Nucl. Instrum. Methods, 92 (1971), 317-23.

613. Prince, M. B. and Polishuk, P. SURVEY OF MATERIALS FOR RADIATION DETECTION AT ELEVATED TEMPERATURES. IEEE Trans. Nucl. Sci., NS-14 (Feb. 1967), 537-543.

614. Prussin, S. G.; Harris, J. S., and Hollander, J. M. APPLICATION OF LITHIUM-DRIFTED GERMANIUM GAMMA-RAY DETECTORS TO NEUTRON ACTIVATION ANALYSIS. Anal. Chem., 37 (Aug. 1965), 1129-32.

A lithium-drifted germanium gamma-ray spectrometer has been utilized for the nondestructive analysis of aluminum by the method of neutron activation.

615. Prussin, S. G.; Harris, J. S., and Hollander, J. M. NONDESTRUC-
     TIVE ACTIVATION ANALYSIS WITH LITHIUM-DRIFTED GERMANIUM
     DETECTORS. pp. 197-199 of "California University, Lawrence
     Radiation Laboratory, Nuclear Chemistry Division. Annual
     Report, 1964." 1965.   UCRL-11828

616. Quittner, P. PEAK AREA DETERMINATION FOR GE(LI) DETECTOR DATA.
     Nucl. Instrum. Sci., 76 (1969), 115-24.

617. Radinov, K. G. and Ren, Li Sam. PREAMPLIFIER FOR GERMANIUM DE-
     TECTORS WITH LOW CAPACITY. Pribory Tekh. Eksper, no. 5
     (Sept. - Oct. 1967), 175-7. (In Russian)

     The noise of a detector-amplifier system depends on both the
     leakage current and the capacity of the detector, as well as on
     the amplifier parameters. For detectors having low capacitances
     the preamplifier tube should be chosen according to the input
     capacity and mutual conductance. A preamplifier is described
     using type 6N23P tubes (c_{in} = 3.2pF) which has an rms noise of
     1.9 keV with Cext = 0.

618. RADIOLOGICAL CHEMISTRY. pp. 3.1-122 of Hanford Radiological
     Science Research and Development. Annual Report for 1964.
     Jan. 1965.   BNWL-36

619. Raeside, D. E. and Wiedenbeck, M. L. MEASUREMENT OF ANGULAR
     CORRELATION SOLID ANGLE CORRECTIONS FOR A COAXIAL GE(LI) DE-
     TECTOR. Nucl. Instrum. Methods, 78 (1970), 331-332.

     Experimental finite solid-angle corrections have been deter-
     mined for a coaxial Ge(Li) detector. A comparison between these
     experimental values and calculated values is shown.

620. Raeside, D. E. PRECISION GAMMA RAY ENERGIES IN THE ENERGY INTER-
     VAL BETWEEN 45 AND 1275 KEV. Nucl. Instrum. Methods, 87
     (1970), 7-11.

     Curved-crystal spectrometers have been utilized to obtain
     measurements of more than forty gamma ray energies in the energy
     interval between 45 and 1275 keV for nuclei with $109 \leq A \leq 193$ in-
     cluding ^{109}Cd, ^{152}Eu, ^{153}Sm, ^{154}Eu, ^{155}Eu, ^{170}Tm, ^{191}Os, ^{165}Eu,
     and ^{193}Os. Many of these energy measurements have sufficiently
     small uncertainties associated with them to make them useful for
     the calibration of Ge(Li) spectra.

621. Ralston, H. Robert and Wilcox, George E. A COMPUTER METHOD OF
     PEAK AREA DETERMINATIONS FROM GE-LI GAMMA SPECTRA. Univer-
     sity of California, Livermore, Lawrence Radiation Laboratory.
     Aug. 1968. 14p.   UCRL-71210; CONF-681003

     Presented at the International Conference on Modern Trends
in Activation Analysis, Gaithersburg, Md.
     A computer method for determining an appropriate base line
(Compton level) from which to begin and end the integration of
a peak area and, also, for determining the amplitude of the Compton continuum at the location of the peak is described.

622. Randle, Keith and Goles, Gordon G. GE(LI) DETECTORS IN THE AC-
     TIVATION ANALYSIS OF GEOLOGICAL SAMPLES. pp. 347-352 of
     "Modern Trends in Activation Analysis." vol. 1. Washing-
     ton, D. C., National Bureau of Standards, 1969.

     A 30 cubic centimeter coaxial lithium-drifted germanium
detector was used to obtain a number of abundances in irradiated
silicic and basic rocks. The resolution of the detector was
about 5.6 keV for the 1333 keV gamma rays of ^{60}Co. Identification of observed gamma-rays based largely on the energies of
the lines and the decay schemes of the isotopes involved. Production and properties of nuclides observable with Ge(Li) detectors are listed. Significant improvement in detector sensitivity was obtained by the use of an anti-Compton shield and by
employing NaI(Tl)-Ge(Li) coincidence measurements.

623. Reidy, J. J. PRELIMINARY RESULTS USING A GERMANIUM LITHIUM-
     DRIFTED DIODE FOR GAMMA RAY SPECTROSCOPY. n.d. 14p.
                                                      TID-21829

     The mounting of the RCA SJGG-1 Li-drifted Ge diode is de-
scribed, and the electronic arrangement is shown graphically.

624. RESEARCH IN NUCLEAR PHYSICS. Progress report no. 16 of Purdue
     Research Foundation, Lafayette, Ind. June 15, 1968. 56p.

     A report giving brief summaries of research on various
nuclear reactions. Results of an investigation on isobaric spin
conservation in the reaction ^{40}Ca (d, α) ^{38}K are presented. The
fabrication and use of Ge(Li) detectors for gamma-gamma angular
correlation measurements are described.

625. Rezanka, I. TIME DEPENDENCE OF COAXIAL-DRIFT PROCESS. <u>Nucl.</u>
     <u>Instrum. Methods</u>, 47 (1967), 181-182.

     Time dependence of intrinsic region formation is derived in
     the approximation of great τ for two limiting cases of coaxial
     drift processes.

626. Ridley, J. D. GE GAMMA-RAY SPECTROMETERS. <u>Nucl. Eng.</u>, 13 (Feb.
     1968), 109-111.

     The principles and construction of a Li-drifted Ge spectrom-
     eters are outlined, including the cryostat design and detector
     mounting. Low noise preamplifier development is described, and
     the interpretation of the gamma spectra obtained is discussed.
     Applications of the spectrometer are reviewed.

627. Ridley, J. D. LARGE VOLUME GERMANIUM GAMMA-RAY SPECTROMETERS.
         pp. 147-151 of "Nucleonic Instrumentation." London, Institu-
         tion of Electrical Engineers, 1968.   CONF-680939

     From the Conference on Nucleonic Instrumentation, Reading,
     England.
     The preparation is described of germanium detectors with
     volumes of 80 cc. The crystals used had a diameter of 46 mm and
     were drifted to a depth of 15 mm.

628. Ridley, J. D. TECHNIQUES EMPLOYED IN THE FABRICATION OF GERMANIUM
         GAMMA-RAY SPECTROMETERS. pp. 29-31 of "Lithium Drifted Ger-
         manium Detectors. Proceedings of a "Panel on the Use of
         Lithium-Drifted Germanium Gamma-Ray Detectors for Research
         in Nuclear Physics, Vienna, 1966." Vienna, International
         Atomic Energy Agency, 1967.

     Techniques employed in the fabrication of germanium gamma-
     ray spectrometer, include fabrication of planar and coaxial Ge(Li)
     detectors at Nuclear Physics Laboratory, University of Oxford.
     U-shaped counters with no insensitive region, small pulse time
     spread and low capacity have been prepared. The choice of mater-
     ial, the details of drifting and clean-up procedure, and the
     difficulties in detector fabrication are presented. The auxili-
     ary equipment needed for the high resolution gamma spectrometers
     with Ge(Li) detectors and characteristics of these spectrometers
     are described.

629. Riepe, G. and Petri, H. GAMMA SPECTROMETRY WITH SEMICONDUCTOR
     DETECTORS. Farnham (Frank C.) Co., Philadelphia, Pa.
     Washington, D. C. August 1967. 14p.
                                     NASA-TT-11202; JUL-308-RB

     Lithium-drifted germanium and silicon semiconducting detec-
     tors were compared with a thallium-activated sodium iodide
     crystal scintillation counter in a multiple channel spectrometer
     for gamma spectrometry. Observed were absorption probability,
     resolution, and time constant of the rising pulse. Obtained
     resolution values proved the germanium-lithium detector superior
     to the scintillation counter; but absorption ability remained be-
     hind the performance of the sodium iodide crystal. The time
     constant of the rising pulse for the germanium sensor was about
     10 times shorter than that for the sodium iodide counter.

630. Rivet, E. J. METHOD FOR COPPER-STAINING GERMANIUM CRYSTALS.
     Nucl. Instrum. Methods, 67 (Jan. 15, 1969), 349-51.

     A method for copper-staining germanium crystals is described.
     This method can be used to show nonuniformities in germanium
     crystals prior to Li drifting, and as a tool for studying the
     diffusion and drifting of Li in germanium.

631. Robinson, D. C. A COMPUTER PROGRAMME FOR THE DETERMINATION OF
     ACCURATE GAMMA RAY LINE INTENSITIES FOR GERMANIUM SPECTRA.
     Nucl. Instrum. Methods, 78 (1970), 120-4.

     A gamma ray analysis programme designed to produce accurate
     peak areas is described. It uses a line shape consisting of two
     Gaussians and an arctangent function together with a cubic back-
     ground to provide a precise fit to full energy gamma ray peaks.
     The final error on the areas after the whole fitting and back-
     ground subtraction process, is only about twice the lower limit
     due to statistical fluctuations. The program is most suitable
     for the analysis of spectra containing only a few well resolved
     peaks.

632. Robinson, R. L.; Stelson, P. H.; McGowan, F. K.; Ford, J. L. C.,
     and Milner, W. T. GAMMA-RAY ENERGIES DETERMINED WITH A
     LITHIUM-DRIFTED GERMANIUM DETECTOR. Nucl. Phys., 74 (1965),
     281-8.

     The energies of gamma rays from about sixty nuclides have
     been determined to within a few tenths of a keV by means of a
     lithium-drifted germanium detector.

633. Robinson, R. L.; Stelson, P. H.; McGowan, F. K.; Ford, J. L. C.,
and Milner, W. T. GAMMA-RAY SPECTROSCOPY WITH A LITHIUM-
DRIFTED GERMANIUM DETECTOR. pp. 116-119 of "Oak Ridge National Laboratory, Tenn., Physics Division. Annual Progress
Report, Dec. 31, 1964-May, 1965." 1965.   ORNL-3778

The energies of gamma rays from 50 nuclides have been measured to within a few tenths of a keV. Gamma-ray doublets in
the decay of ^{102}Rh and ^{106}Rh and in the Coulomb excitation of
^{77}Se are reported.

634. Rodda, J. L., Jr.; Macklin, R. L., and Gibbons, J. H. RESPONSE
OF 25 cm^3 GE(LI) DETECTOR TO NEUTRONS: SHIELDING FACTORS.
Nucl. Instrum. Methods, 74 (1969), 224-228.

The response of a large Ge(Li) detector to fast and slow
neutrons was investigated. Characteristic gamma-ray pulse
height distributions from a few hundred keV to over 1 MeV were
obtained and prominent lines were identified. The effectiveness
of an anticoincidence plastic scintillator shield, combined with
a radiationless neutron absorber and a lead gamma absorber, was
evaluated and found to be in good agreement with the calculated
response.

635. Rodionov, K. G. and Ren, Li Sam. PREAMPLIFIER FOR GERMANIUM DE-
TECTORS HAVING A LOW CAPACITY. Instrum. and Exper. Tech.,
no. 5 (1967), 1151-3.

The noise of a detector-amplifier system depends on both the
leakage current and the capacity of the detector, as well as on
the amplifier parameters. For detectors having low capacitances
the preamplifier tube should be chosen according to the input
capacity and mutual conductance. A preamplifier is described
using type 6N23P tubes (C_{in}=3.2pf) which has an rms noise of
1.9 keV with C_{ext}=0.

636. Romantschuk, H.; Kauranen, P.; Rahola, Tua; Hattula, Tuulikki,
and Miettinen, J. K. GAMMA-SPECTROMETRIC IDENTIFICATION OF
FALLOUT NUCLIDES WITH A GE(LI)-DETECTOR. Suomen Kem., B-39
(1966), 182-6.

A large rainwater sample was collected about one month after
the detonation of the Chinese nuclear bomb on May 9, 1966. The
sample was evaporated to dryness and the residue γ-counted
using a Ge(Li)-semiconductor and a multichannel analyzer. Several
fallout nuclides were identified solely on the basis of their

γ-energies. [140]Ba was also determined by radiochemical separation.

637. Rotolante, Ralph Allen. GE(LI) DETECTOR FABRICATION, A GE(LI)-
      GE(LI) MULTIPARAMETER COINCIDENCE EXPERIMENT AND THE LEVELS
      OF [185]Re. Nashville, Vanderbilt University. 1970. 188p.
      Thesis

      Equipment and techniques for the fabrication of lithium-drifted
      germanium (Ge(Li)) detectors are described. Also discussed in de-
      tail are the operating characteristics of solid-state detectors,
      with particular emphasis given to the timing characteristics of
      such diodes. The feasibility of performing low-energy, quanti-
      tative coincidence experiments with state-of-the-art multi-
      channel analyzers and Ge(Li) detectors is studied. The results
      were extended to very low energies (~50 keV). The coincidence
      study was done in conjunction with the level-scheme study of
      the isotope [185]Re, and quantitative results regarding transi-
      tion intensities and their interpretation are presented. Finally,
      a discussion of the structure of the various excited levels is
      given.

638. Rotolante, Ralph Allen; Albridge, R. G., and Fleming, R. M.
      SIMPLE METHOD OF DRIFT CONTROL FOR GE(LI) DETECTORS. Nucl.
      Instrum. Methods, 65 (Nov 1968), 235-6.

639. Rouse, Robert Lindsay. IMPROVEMENTS RELATING TO SOLID STATE RA-
      DIATION DETECTORS. Sept. 29, 1965. Brit. Patent 1,006,233

      A method of manufacturing p-i-n junction semiconductor radia-
      tion detectors includes the steps of treating a semiconductor al-
      loy of Si and Ge (at least 10% Ge) by a lithium ion drift process
      at a high temperature to form a p-n junction, applying a reverse
      electrical bias to the junction at lower temperature to increase
      the thickness of the depletion zone and form a p-i-n junction and
      connecting electrical leads to the device for connection to ex-
      ternal circuitry. Appropriate circuitry may include a biasing
      source connected to apply a reverse bias to the p-i-n detector and
      a pulse height analyzer for measuring the output. The device may
      be cooled below ambient temperature.

640. Roush, M. L. and Connors, P. I. EFFECT OF UNDEPLETED REGION UPON
      DOPPLER-BROADENED LINE SHAPE FOR GE(LI) DETECTORS. Nucl.
      Instrum. Methods, 70 (Apr. 15, 1969), 218-20.

      Line shapes observed with a coaxial Ge(Li) detector are af-

fected by the presence of the undepleted core when the γ-ray
energy varies with position across the face of the detector.

641. Roussel, Louis. STABILIZED CONSTANT POWER SUPPLY. Commissariat
     a l'Energie Atomique, Saclay, France. Centre d'Etude
     Nucleaires. Oct. 1968. 123p. Thesis. (In French)
                                        Also report CEA-R-3631

     The study and realization of a stabilized power supply vari-
     able from 5 to 100 watts are described. In order to realize a
     constant power drift of lithium compensated diodes, a 1 percent
     precision of regulation and a response time less than 1 sec was
     investigated. Recent components like Hall multiplicator and
     integrated amplifiers give this possibility and it is easy to
     use permitable circuits.

642. Routti, Jorma T. SAMPO: A FORTRAN IV PROGRAM FOR COMPUTER ANAL-
     YSIS OF GAMMA SPECTRA FROM GE(LI) DETECTORS, AND FOR OTHER
     SPECTRA WITH PEAKS. University of California, Berkeley.
     Lawrence Radiation Laboratory. Oct. 20, 1969. 34p.
                                        UCRL-19452

     SAMPO is a Fortran IV program written to perform the data-
     reduction analysis described by J. T. Routti and S. G. Prussin
     in Photopeak Method for the Computer Analysis of Gamma-Ray Spec-
     tra from Semiconductor Detectors, Nuclear Instruments and Methods
     72, 125-142 (1969). The mathematical methods and their coding
     are briefly described. Instructions for using the program and
     for preparing input data are given and the optimal strategies
     for running the code are discussed.

643. Ruge, I.; Eichinger, P., and Koepp, F. FABRICATION OF LITHIUM-
     DRIFTED GERMANIUM PIN DETECTORS FOR HIGH-RESOLUTION GAMMA
     SPECTROSCOPY. Kerntechnik, 8 (July 1966), 298-300.

     The principle of the lithium-drift is given and the germanium-
     lithium system is described. The drift procedures are outlined
     with emphasis on the difficulties of preparation. A method for
     the preparation of germanium p-i-n detectors for γ-spectroscopy
     is given.

644. Ryndina, E. Z. LARGE VOLUME GE(LI)-DETECTORS FOR GAMMA-RAY
     SPECTROSCOPY. At. Energy. (USSR), 27 (July 1969), 64-66.
     (In Russian)

Construction and performance of a coaxial high-sensitivity large volume (30 cm^3) γ-spectrometer are described.

645. Ryvkin, S. M.; Mateev, O. A.; Strokan, N. B., and Khusianov, A. Kh. GERMANIUM GAMMA-QUANTUM COUNTER WITH RADIATIVE DEFECTS. Zh. Tekhn. Fiz., 34 (Aug. 1964), 1535-37. (In Russian)

Descriptions are given of germanium γ counters with deep level compensation produced by radiation defects. The energy spectra of ^{137}Cs recorded by such counters indicate a resolving power of ~3% with an energy of pair formation 3.0 ± 0.1 ev.

646. Ryvkin, S. M.; Makovsky, L. L.; Strokan, N. B.; Subashieva, V. P., and Khusainov, A. Kh. PREPARATION OF STABLE COUNTERS BY MEANS OF COMPENSATING GERMANIUM WITH RADIATION PRODUCED STRUCTURAL DEFECTS. IEEE Trans. Nucl. Sci., NS-15, no. 3 (June 1968), 226-231.

Gamma-quantum Ge-counters using germanium compensated by "deep" levels offer considerable advantages as compared with Ge(Li) counters. Characteristics features of obtaining a strong electric field in p-i-n structures under conductivity compensation by "deep" levels have been discussed.

647. Ryvkin, S. M.; Strokan, N. B.; Subashieva, V. P.; Tisnek, N. I., and Khusainov, A. Kh. ROLE OF ELECTRIC FIELD INHOMOGENEITIES IN SEMICONDUCTOR NUCLEAR RADIATION COUNTERS. Sov. Phys. Semicond. (Engl. Transl.), 4 (Jan. 1971), 1107-12.

Translated from Fiz. Tekh. Poluprov., 4 (Jul. 1970), 1303-10. An investigation was made of the influence of structure defects in n-type germanium on the energy resolution of counters prepared by the "cold" doping method. The samples were compensated by acceptor centers generated by irradiation with ^{60}Co γ-quanta. The crystal blanks were scanned with a light probe and with narrow beam particles. The topography of the signals generated by these probes reflected the degree of inhomogeneity of the original and irradiated germanium. Structure defects, which governed the breakdown voltage of p/n junctions, could be revealed by chemical etching of the surfaces of the disk-shaped blanks. An analysis of the nature of the amplitude spectra of the counters made it possible to estimate the dimensions and configurations of inhomogeneous regions and to separate the influence of inhomogeneous distributions of the electric field and carrier lifetime.

648. Ryvkin, S. M.; Matveev, O. A.; Strokan, N. B., and Khusainov,
     A. Kh. SPECTROMETRIC GAMMA-QUANTUM COUNTER BASED ON GER-
     MANIUM WITH RADIATION DEFECTS. Dokl. Akad. Nauk SSSR, 165
     (Nov. 21, 1965), 548-50. (In Russian)

     The design and operating characteristics of semiconductor
     gamma-counters based on germanium with radiation defects pro-
     duced by gamma rays of ^{60}Co are discussed. The counters were
     found to possess features superior to those of lithium-doped de-
     tectors with respect to amplitude resolution.

649. Sadokhin, I. P. and Lashchuk, A. I. A GE(LI) DETECTOR. Pribory
     i Tekhnika Eksp., no. 4 (1968), 799-800.

     An anticoincidence gamma spectrometer using a Ge(Li) detector
     1.6 cm^3 in volume is described. The resolution of the spectrom-
     eter for an energy of 122 keV (Co^{57}) is 4.5 keV. The Compton
     distribution is reduced to between 50 and 70%.

650. Sakai, Eiji and Fowler, I. L. CHARACTERISTICS OF COAXIAL GE(LI)
     DETECTORS THROUGH WARM-UP AND RE-DRIFT CYCLES. IEEE Trans.
     Nucl. Sci., NS-15, no. 3 (June 1968), 327-31.
                                          Also report CONF-680207

     From the 11th Scintillation and Semiconductor Counter Symposi-
     um, Washington, D. C.
     Studies have been made of several cylindrical coaxial Ge(Li)
     detectors through various cycles of warming to room temperature
     and subsequent cleanup re-drift all in high vacuum. Conventional
     double-open-end detectors with an n+-layer on the outside and a
     p-type core were used. Using the gamma-ray scanning technique
     it was observed that when not fully cleaned up (capacitance de-
     creasing with increasing voltage) the i-layer thickness increased
     from the n-layer towards the p-core as voltage was increased,
     after a warm-up period, part of the i-layer nearest the core was
     lost with reduction efficiency (18% at 500V for one detector after
     an 8 hour warm-up period), and cleanup re-drift following a warm-
     up period restored the original characteristics but the clean-up
     time varied with the crystal from which the detector was made.

651. Sakai, Eiji. CHARGE COLLECTION IN COAXIAL GE(LI) DETECTORS. IEEE
     Trans. Nucl. Sci., NS-15, no. 3 (June 1968), 310-20.
                                          Also report CONF-680207

     From the 11th Scintillation and Semiconductor Counter Symposi-
     um, Washington, D. C.

Cylindrical double-open-end coaxial Ge(Li) detectors have been studied at 77°K using a small-area (circular) collimated beam of ^{137}Cs gamma-rays. The beam was aimed parallel to the detector axis and moved across a diameter. On different detectors, different trapping effects were observed; one detector gave better resolution and a greater pulse height when irradiated near the p-type core, another gave best performance when irradiated near the outer n layer. The observed pulse shapes for different positions of irradiation showed general agreement with calculated pulse shapes based on an assumed r^{-1} variation of electric field across the i-region.

652. Sakai, Eiji; Suzuki, Toshio; Gotoh, Hiroshi , and Sekiguchi, No-
      butada. A COAXIAL GE(LI) GAMMA-RAY SPECTROMETER. J. Nucl.
      Sci. Technol. (Tokyo), 3 (Dec. 1966), 534-8.

A single-open-end coaxial type Ge(Li) gamma-ray detector with an effective volume of 8.7 cm^3 and a capacitance of 35 pF was made by Li-drifting for 440 hr at an applied voltage of 140V at 40°C in thermomodule-controlled constant-temperature water coolant. The fwhm energy resolution with an E810F preamplifier was 4.2 and 8.0 keV for gamma rays of 0.122 and 1.173 MeV, respectively. The pulse height distributions of gamma rays from an irradiated natural U sample and ^{105}Rr isotope were measured.

653. Sakai, Eiji; McMath, T. A., and Franks, R. G. FURTHER CHARGE
      COLLECTION STUDIES IN COAXIAL GE(LI) DETECTORS. IEEE Trans.
      Nucl. Sci., NS-16, no. 1 (Feb. 1969), 68-74.

654. Sakai, Eiji; Malm, H. L., and Fowler, I. L. PERFORMANCE OF GER-
      MANIUM (LITHIUM) DETECTORS OVER A WIDE TEMPERATURE RANGE.
      pp. 101-120 of "Semiconductor Nuclear-Particle Detectors and
      Circuits." Brown, W. L., ed. Washington, D. C., National
      Academy of Sciences, 1969.   Also report CONF-670520

From the Conference on Semiconductor Nuclear-Particle Detectors and Circuits, Gatlinburg, Tenn.
A study was made of the characteristics of Ge(Li) gamma spectrometers over the temperature range 5 to 220°K. Properties studied included resolution, leakage current, transit time, and maximum and minimum operating temperature.

655. Sakai, Eiji and Malm, H. L. PERFORMANCE OF GE(LI) DETECTORS IN
      THE TEMPERATURE RANGE 5.0 TO 170°K. Appl. Phys. Lett., 10
      (May 15, 1967), 268-70.

The properties of three Ge(Li) detectors as gamma-ray spectrometers were measured over the temperature range 5.0 to 170°K.
These detectors, selected to show a range of charge-trapping
effects, all exhibited predominantly electron trapping, and
optimum resolution occurred between 20 and 30°K were electron-
trapping effects were a minimum.

656. Sakai, Eiji and McMath, T. A. A RISETIME SPECTROMETER FOR
STUDYING PULSE SHAPES IN GE(LI) GAMMA-RAY DETECTORS. Nucl.
Instrum. Methods, 64 (1968), 132-140.

A risetime spectrometer has been developed to study the shape
and distribution of shape of gamma ray pulses from Ge(Li) detectors. This instrument employs a time to amplitude converter, a
multichannel pulse height analyzer, and two discriminators to
measure the distribution of times required for full energy gamma
ray pulses to reach a predetermined fraction of full pulse
height, using the risetime spectrometer and collimated beam techniques, pulse shapes from a planar and a coaxial detector were
determined as a function of the position of irradiation in the
detector.

657. Sakai, Eiji. SEMICONDUCTOR GAMMA-RAY DETECTORS. Oyo Butsuri,
38 (Jan. 1969), 2-19. (In Japanese)

Recent progresses on studies of charge collection process in
Ge(Li) detectors and problems in their manufacturing are described. Topics such as radiation damage to Ge(Li) detectors,
well-type Ge(Li) detectors, double-diode Compton suppression
gamma-ray spectrometers, and CdTe detectors are also reviewed.

658. Sakai, Eiji. SLOW PULSES FROM GERMANIUM DETECTORS. IEEE Trans.
Nucl. Sci., NS-18, no. 1 (Feb. 1971), 208-218.

Slow pulses from various types of Ge(Li) detectors, i.e. a
planar, a coaxial, a thin p+-contact planar, and a back-to-back
double planar detector, have been studied by using a gamma ray
scanning technique and a risetime selector.

659. Sakai, Eiji. TEMPERATURE DEPENDENCE OF THE AVERAGE ENERGY PER
ELECTRON-HOLE PAIR FOR GAMMA-RAYS IN GE(LI) DETECTORS. IEEE
Trans. Nucl. Sci., NS-15, no. 1 (Jan. 1968), 432-437.

660. Salmon, L. RECENT ADVANCES IN THE ANALYSIS OF PULSE HEIGHT DIS-
     TRIBUTIONS IN GAMMA-RAY SPECTROMETRY. pp. 159-65 of "Con-
     ference on Nucleonic Instrumentation." London, Institution
     of Electrical Engineers, 1968.   CONF-680939

     The present use of gamma-ray spectrometry using both sodium
     iodide and lithium drifted germanium detectors has extended into
     many fields of physics and chemistry. In particular, routine
     analysis of radionuclide mixtures are performed by this technique
     far more rapidly and cheaply than by laborious conventional radi-
     ochemical methods.

661. Santhanam, S. and Monaro, S. A WELL-TYPE GE(LI) DETECTOR FOR
     SUM-COINCIDENCE STUDIES. Nucl. Instrum. Methods, 76 (1969),
     322-27.

     The feasibility and usefulness of a coaxial Ge(Li) spectrom-
     eter as a well-type detector for sum spectra of coincident gamma
     rays are demonstrated. The value of the well-type geometry has
     previously been established in nuclear gamma spectroscopy with
     the use of NaI(Tl) detectors. In this work a 16 cm^3 hollow
     cylindrical germanium detector has been made and mounted in such a
     way that access to the denter of the detector is achieved from
     outside the cryostat. The detector is a fully depleted structure
     in which a hole of 1 cm diameter was ultrasonically cut, and a
     non-injecting contact applied to the inside surface.

662. Sarantites, D. G. and Gronemeyer, S. A GE(LI)-GE(LI) STUDY OF
     THE DECAY OF 2.4h ^{66}Ge. Nuclear Phys. (Netherlands), A130,
     no. 1 (June 2, 1969), 97-111.

     The level structure of ^{66}Ge was investigated from decay of
     the 2.4h ^{66}Ge. From singles gamma ray energy and intensity
     measurements employing a high resolution Ge(Li) detector, from
     coincidence relationships determined in γ-γ coincidence measure-
     ments employing two Ge(Li) detectors and from $\beta\pm\gamma$ coincidence
     measurements, it was established that levels at 43.5, 108.9,
     233.9, 318.7, 422.8, 514.3, 536.5, 706.3, and 866.1 keV are popu-
     lated in the decay of 2.4h ^{66}Ge. A possible level at 639.9 is
     also considered. From present log ft values and relative gamma
     ray intensities many J^{π} values were assigned.

663. Saunders, E. W. and Maxwell, C. J. PARALLELING PLANAR GE(LI)
     DETECTORS FOR COUNTING LARGE VOLUME BIOLOGICAL SAMPLES.
     University of California, Livermore, Lawrence Radiation
     Laboratory. Oct. 25, 1967. 29p.

CONF-671026-5; UCRL-70545

Presented at the 14th Nuclear Science Symposium, Los Angeles, California.

Four planar Ge(Li) detectors have been operated in an array, inparallel electrically, and connected to a single preamplifier. The total active volume is 28 cm. The photopeak efficiency of the four-detector array at 1.33 MeV is about 4.8 times greater than the efficiency of one 7 cm^3 detector. The resolution is 2.90 keV FWHM at 1.33 MeV and 2.01 keV FWHM at 122 keV. Some misalignment of the individual detector photopeaks is observed, probably caused by contact resistance. A technique which allows alignment of the peaks is presented.

664. Sayre, Edward. REFINEMENT IN METHODS OF NEUTRON ACTIVATION ANAL-
     YSIS OF ANCIENT GLASS OBJECTS THROUGH THE USE OF LITHIUM
     DRIFTED GERMANIUM DIODE COUNTERS. Brookhaven National Lab-
     oratory, Upton, New York. 1965. 28p.
                          BNL-9614; Also report CONF-650640-2

From the 7th International Glass Congress, Brussels.

With new lithium drifted germanium diode solid state detectors, which record photopeaks with nearly ten times the resolution of large crystal scintillation detectors, it was possible
to determine non-destructively the concentrations of some rare
earth oxides in ancient glasses.

665. Sayres, A. R. and Baicker, J. A. THE ALL GERMANIUM ANTI-COMPTON
     SPECTROMETER. IEEE Trans. Nucl. Sci., NS-15, no. 3 (June
     1968), 393-396.

A two crystal Ge(Li) detector system has been developed for
use as an anti-Compton gamma-ray spectrometer. The basic components of this system are a pair of Ge(Li) detectors mounted
in tandem.

666. Schell, K. J. and Nienhuis, K. A STUDY OF TRAPPING CENTRES IN
     GE(LI) DETECTORS. pp. 37-54 of the "Proceedings of the
     Meeting on Special Techniques and Materials for Semiconductor
     Detectors, Ispra, Italy, 1968." June 1969.   EUR-4269

A study has been made on the influence of trapping centers
in Ge(Li) detectors. The relation between the depth of the
trapping center and the resolution of the detector has been investigated. Only a small range of energy levels of the trapping
centers is responsible for trapping effects in the detector.

One of the few impurities that fulfills all the requirements is
silver. This assumption is strongly supported by experiments.

667. Schmidt, C. T. LITHIUM-DRIFTED GERMANIUM GAMMA DETECTOR.
         pp. 3-6 of the "Hazards Control, Quarterly Report, no. 24."
         Lawrence Radiation Laboratory, University of California,
         Livermore. March 1966.

668. Schmidt-Whitley, R. D. PULSE SHAPE CHARACTERISTICS OF A GE(LI)
         DETECTOR STUDIED WITH A COLLIMATED GAMMA-RAY BEAM. Nucl.
         Instrum. Methods, 70 (Apr. 15, 1969), 227-32.

A rectangular Ge(Li) detector drifted from five sides is
investigated with a collimated beam of gamma rays. The pulse
rise shape distributions and their influence on coincidence
timing are discussed.

669. Schroeder, Gerald L.; Kraner, Hobart W., and Evans, Robley D.
         THE APPLICATION OF LITHIUM-DRIFTED GERMANIUM DETECTORS TO
         NEUTRON ACTIVATION ANALYSIS-STUDIES IN GEOCHEMISTRY AND OF
         15TH CENTURY PRINTING. Trans. Amer. Nucl. Soc., 8 (Nov. 1965),
         327.

670. Schroeder, Gerald L.; Kraner, Hobart W., and Evans, Robley D.
         LITHIUM-DRIFTED GERMANIUM DETECTORS: APPLICATION TO NEUTRON-
         ACTIVATION ANALYSIS. Science, 151 (Feb. 18, 1966), 815-817.

Lithium-drifted germanium detectors for high-resolution gamma-
ray spectroscopy reduce the need for wet chemistry in neutron-
activation analysis. Problems in fields as diverse as geo-
chemistry and the history of 15th century printing are shown to
be susceptible to this analytic technique.

671. Schueler, W. A. NOVEL LITHIUM DRIFT CONTROL METHOD FOR SILICON
         AND GERMANIUM RADIATION DETECTORS. Rev. Sci. Instrum., 38
         (Apr. 1967), 539-41.

A method and an apparatus are described which allows a nondestructive indication of the progress of the drifting operation.
While drifting, pulsed light is shone onto the face opposite the
lithium diffused layer of the device. The photosensitivity of
the detector is used to obtain a signal whose amplitude is
inversely proportional to the window thickness, i.e., the thickness of the undrifted region. A charge sensitive amplifier and

a peak rectifier convert the pulses into a dc signal which is used for window thickness indication.

672. Scott, H. L.; Horoshko, R. N., and Van Patter, D. M. REACTION INDEPENDENT ANALYSIS OF GAMMA RAY ANGULAR DISTRIBUTIONS FROM GE(LI)-DETECTOR MEASUREMENTS. Nucl. Instrum. Methods, 70 (May 1, 1969), 320-4.

Reaction-independent analysis of particle gamma-angular correlations from aligned nuclei involves particle detection near 0° or 180°m and is generally limited by the energy resolution of the particle counter if a NaI(Tl) counter is used to detect the gamma-rays, and by the relatively low coincidence rates if a Ge(Li) counter is used.

673. Seagondollar, L. W. PROGRESS REPORT (ON NUCLEAR PHYSICS), NOVEMBER 1, 1967. North Carolina State University, Raleigh. 1967. 61p.   ORO-3624-1

Techniques were developed for fabricating Ge(Li) detectors, and a three-crystal pair spectrometer using scintillation and semiconductor detectors was designed.

674. Segel, R. E.; Siemssem, R. H.; Blaugrund, A. E.; Baker, S. I., and Morrison, G. C. LIFETIME MEASUREMENTS BY THE DOPPLER-SHIFT METHOD. pp. 9-10 of the "Argonne National Laboratory, Ill., Physics Division Report, April 1966." 1967.

675. Selin, E.; Arnell, S. E., and Almen, O. ON THE PROPERTIES OF ELECTROMAGNETICALLY SEPARATED NOBLE GAS TARGETS FOR NUCLEAR REACTION STUDIES. Nucl. Instrum. Methods, 56 (1967), 218-228.

In the present investigation backing materials for electromagnetically separated noble gas targets are discussed. Ion beams of ^{22}Ne and ^{40}Ar in the energy range 2 - 75 keV have been shot into the common backing materials tantalum and molybdenum, and the distributions of the collected noble gas atoms in saturated backings have been studied by means of (p, γ)-reactions. Tables of experimental half-widths of the distributions are given in μg cm^2. Some measurements have been made in order to elucidate the process.

676. Senftle, F. E.; Evans, A. G.; Duffey, D., and Wiggins, P. F.
     CONSTRUCTION MATERIALS FOR NEUTRON CAPTURE-GAMMA-RAY MEASURE-
     MENT ASSEMBLY USING ^{252}Cf. <u>Nucl. Technol.</u>, 10 (Feb. 1971),
     204-10.

     To measure the neutron capture gamma ray spectrum of a sam-
     ple using neutrons from a multimicrogram ^{252}Cf source, the
     Ge(Li) detector must be in a position to "see" the source. Thus,
     the encapsulation, moderating and shielding materials all become
     potential sources of spectral interference. An interference
     parameter is calculated for a number of elements generally found
     in the associated source and detector hardware. Using this param-
     eter, other construction materials can also be chosen to mini-
     mize spectral interference in measuring neutron capture gamma
     ray spectra.

677. Sever, Y. A. COMPTON-REJECTION GERMANIUM SPECTROMETER. <u>Nucl.
     Instrum. Methods</u>, 33 (Mar. 1965), 347-8.

     When lithium-drifted germanium detectors are used for gamma
     spectrometry instead of NaI crystals, the advantage of higher
     resolution is much dampened by the height (relative to the photo-
     peak) and sharpness of Compton edges. The danger of misinter-
     pretation of a Compton edge as a photopeak in a composite spec-
     trum is stronger in this case. A 1 cm^2 x 77 mm germanium de-
     tector with a 5 mm thick lithium-drifted layer that was guarded
     by a 108 mm diameter NaI(Tl) well counter with wall thickness
     of 31 mm is described. The spectrum of ^{137}Cs with and without
     Compton rejection is illustrated.

678. Sever, Y. A. PRODUCTION OF GERMANIUM LITHIUM DRIFTED DETECTORS
     AT THE SOREQ NUCLEAR RESEARCH CENTER. Israel Atomic Energy
     Commission. Yavne. Soreq Nucl. Research Center. Jan. 1968.
     10p.   IA-1153

     The advantages of laboratory-constructed lithium-drifted
     germanium detectors are discussed and some features of detector
     production are described. Data on the performance of a labora-
     tory-constructed detector are given.

679. Sher, Alvin H. CARRIER TRAPPING IN GE(LI) DETECTORS. <u>IEEE Trans.
     Nucl. Sci.</u>, NS-18, no. 1 (Feb. 1971), 175-183.

680. Sher, Alvin H. FABRICATION AND APPLICATION OF SEMICONDUCTOR RA-
     DIATION DETECTORS IN HIGH RESOLUTION NUCLEAR DECAY STUDIES.

Simon Frazer University. n. d. Thesis

A discussion of the factors which determine performance of
lithium-drifted semiconductor radiation detectors, particularly
lithium-drifted Ge detectors for gamma ray spectroscopy is
presented.

681. Sher, Alvin H. and Pate, B. D. GE(LI) DETECTORS WITH UNUSUAL
     OPERATING CHARACTERISTICS. Nucl. Instrum. Methods, 53
     (1967), 339-340.

Ge(Li) detectors of up to 1.5 cm^3 active volume have been
operated at average biases of 70 v with good charge collection
and resolution. The possible effect of detector shape on performance is discussed.

682. Sher, Alvin H. and Coleman, J. A. LITHIUM DRIFTABILITY IN DE-
     TECTOR-GRADE GERMANIUM. IEEE Trans. Nucl. Sci., NS-17,
     no. 3 (June 1970), 125-9.   Also report CONF-700301

The behaviour of many germanium crystals during the lithium-
ion drift in terms of compensated depth W as a function of drift
t, does not follow the prediction of the equation: $W=(2\mu_{Li}Vt)1/2$.
Modification of the lithium velocity expression to include that
loss of mobile lithium ions during the drift yields the expression: $W=(\mu_{Li}Vt(1-e^{-2t}/\tau))1/2$, which seems to describe satisfac-
torily the experimental results. In addition to the loss mechanism, the effect of oxygen on lithium-ion drift mobility is
considered.

683. Sher, Alvin H. NOMOGRAPHS FOR USE IN THE FABRICATION AND TESTING
     OF GE(LI) DETECTORS. Washington, D. C., National Bureau of
     Standards. Aug. 1970. 18p.

Six nomographs which can facilitate the fabrication and
testing of lithium-drifted germanium gamma-ray detectors (Ge(Li)
detectors) have been constructed which relate the following
parameters: (1) time, temperature, applied bias, and drifted
depth; (2) lithium mobility, crystal resistivity, and oxygen
concentration; (3) area, capacitance, and drifted depth for
planar Ge(Li) detectors; (4) drifted depth, length and capacitance for coaxial Ge(Li) detectors; (5) total spectral resolution, system noise, and detector resolution; and, (6) detector
resolution, gamma-ray energy, and effective Fano factor. The
use of these nomographs is described and illustrative examples
are given.

684. Sher, Alvin H. and Keery, W. J.   VARIATION OF THE EFFECTIVE FANO
        FACTOR IN A GE(LI) DETECTOR.   IEEE Trans. Nucl. Sci., NS-17,
        no. 1 (Feb. 1970), 39-43.   Also report CONF-691017-Pt.1

        Variations in the value of the effective Fano Factor, F',
        have been observed in Ge(Li) detectors using a collimated beam
        of gamma-rays to irradiate selected areas in the sensitive region
        between the n+ and p-contacts. For regions near the n+ contact
        where trapping of electrons is minimized, F' was found to be
        statistically less than 0.11.

685. Sherman, N. K. and Horowitz, Y. S.   SEPARATION OF PARTICLES AT
        INTERMEDIATE ENERGIES IN A GERMANIUM RANGE-TELESCOPE SPEC-
        TROMETER.   Nucl. Instrum. Methods, 56 (1967), 106-108.

        At intermediate energies, clean proton scattering spectra
        and heavy particle reaction spectra can be obtained simultaneous-
        ly using a two detector range telescope. The method is based on
        the large difference between the range of protons and the range
        of all heavier particles of comparable energy. Its application
        at 100 MeV using germanium detectors to study ^{12}C and ^{9}Be is
        described.

686. Shipp, R. L. and Brashear, H. R.   LITHIUM-DRIFTED GERMANIUM DE-
        TECTOR DEVELOPMENT.   p. 31 of "Oak Ridge National Laboratory.
        Instrumentation and Control Division, Annual Progress Report,
        Sept. 1, 1967." 1967.   ORNL-4219

687. Shirley, David A.   APPLICATIONS OF GERMANIUM GAMMA-RAY DETECTORS.
        Nucleonics, 23 (Mar. 1965), 62-6.   Also report UCRL-11865

        The performance of lithium-drifted germanium detectors in
        gamma spectrometry is discussed in relation to that of NaI(Tl)
        devices, showing the superior resolution and speed of the former
        applications of germanium detectors to gamma spectrum analysis;
        coincidence and angular-correlation measurements, moessbauer
        spectroscopy; and studies of nuclear orientation, conversion
        coefficients, neutron-capture gamma rays and meson (μ) x-rays are
        then considered.

688. Siffert, P. and Regal, R.   LOW NOISE CHARGE PREAMPLIFIER FOR
        LITHIUM-COMPENSATED GERMANIUM DETECTORS OF LARGE SENSITIVE
        VOLUME.   Rev. Phys. Appl. (Suppl. J. Phys.), 3 (June 1968),
        107-110. (In French)

A charge preamplifier associated with a germanium lithium gamma-ray detector of large sensitive volumes (85 cm^3) is described. This preamplifier uses, as input element, a field effect tetrode transistor. When the transistor is cooled to 150° K and used with the gamma counter, a 4.3 keV FWHM for a 1.33 MeV gamma-ray may be obtained.

689. Simmons, S. Oliver. CONSTRUCTION OF A THREE CRYSTAL SPECTROMETER AND A STUDY OF THE DECAY OF ^{90}Nb. Iowa State University. October 1970. 72p. Thesis ◆   Also report IS-T-389

A gamma ray spectrometer consisting of a Ge(Li) detector partially surrounded by two NaI(Tl) scintillators has been constructed. The instrument has two modes of operation, the pair spectrometer mode for analyzing high energy gamma rays (above 1.022 MeV) and the Compton suppression mode for studying weak gamma rays in the presence of intense higher energy background radiation. The decay of ^{90}Nb to levels in ^{90}Zr was investigated using the three crystal spectrometer in the Compton suppression mode to reduce background from the intense 2186 and 2319 keV transitions. Energies and absolute intensities of thirteen gamma ray transitions were determined. Using the K-conversion electron intensities from the work of other investigators and the gamma ray intensities from this work, conversion coefficients for the transitions were calculated. Comparison with theoretical values enabled deduction of the multipolarities of most of the transitions. The conventional shell model description of the levels of ^{90}Zr in terms of two protons in (p 1/2) and (g 9/2) orbitals is inconsistent with the observation of a strong E1 transition between the first 6+ and 5- levels. A calculation is presented which shows that this can be explained by mixing a 10-2% component of (G 9/2) (h 11/2) into the wave function for the 5-state.

690. Sklavenitis, Laodames. LA SPECTROMETRIE GAMMA FINE PAR JONCTION P-I-N. APPLICATION A L'ANALYSE PAR ACTIVATION. (FINE GAMMA SPECTROMETRY USING P-I-N JUNCTIONS. APPLICATION TO ACTIVATION ANALYSIS). Commissariat a l'Energie Atomique, Saclay, France. Centre d'Etudes Nucleaires. Dec. 1967. 53p. (In French) Thesis   Also report CEA-Bib-110

The author reviews the properties of solid detectors, their functioning and the various types existing at present. He describes further in more detail the gamma spectrometer using a lithium-drifted germanium detector and its application in activation analysis.

691. Slivinsky, V. W. and Ebert, P. J. EFFICIENCY CALIBRATION OF A
     GE(LI) DETECTOR FROM 8 TO 98 KEV. Nucl. Instrum. Methods,
     71 (1969), 346-8.

     The efficiency of a Ge(Li) detector was measured using x-ray
     machines in the energy region from 8 to 98 keV by comparing the
     number of fluorescent x-rays counted by the Ge(Li) spectrometer
     with that by a totally absorbing NaI(Tl) spectrometer. The
     escape peak-to-photopeak ratios as measured are roughly 35%
     higher than those of other published results. Monte Carlo cal-
     culations of the detector efficiency and escape fraction at
     different photon energies compare favorably with experimental
     data; these should be useful for other workers using a highly
     collimated Ge(Li) detector, which approximates semi-infinite
     geometry.

692. Smith, K. F. and Cline, J. E. A LOW-NOISE CHARGED SENSITIVE
     PREAMPLIFIER FOR SEMICONDUCTOR DETECTORS USING PARALLEL
     FIELD-EFFECT-TRANSISTORS. IEEE Trans. Nucl. Sci., NS-13,
     no. 3 (June 1966), 468-476.

693. Sodd, V. J.; Wellman, H. N., and Branson, B. M.   [123]I THYROID
     MEASUREMENTS WITH A GE(LI) DETECTOR. J. Nucl. Med., 10
     (Mar. 1969), 136-9.

     The high resolution of the Ge(Li) semiconductor detector and
     its reasonably good counting efficiency for low-energy gamma rays
     suggest its use in nuclear medicine. A Ge(Li) detector with a
     sensitive volume of 4.2 cm^3 was used to measure human thyroid
     uptake of [123]I. A Lucite neck phantom was used to calibrate
     this detector system for varying thyroid depths using the P/S
     ratio technique. The percent uptake measured in two subjects
     using the calibrated Ge(Li) detector agreed with measurements
     made with a dual NaI(Tl) detector.

694. Sonrel, C.; Perrin, P.; Maurel, M.. and Mougin, A. EFFICIENCY AND
     TIME RESOLUTION FOR LARGE GE(LI) GAMMA DETECTORS USED IN FAST
     NEUTRON EXPERIMENTS. Commissariat a l'Energie Atomique,
     Grenoble, (France). Centre d'Etudes Nucleaires. 1969. 21p.
                                                          CEA-CONF-1432

     The intrinsic efficiency, energy and time resolution for the
     range of 5 to 17 MeV have been measured for some large (20 to
     120 cm^3) Ge(Li) detectors. The possibility of (n, γ) and
     (n, n', γ) experiments has been checked using a multiparametric
     data system.

695. Sovka, Jerry A. and Rasmussen, Norman C. NONDESTRUCTIVE ANALYSIS
     OF IRRADIATED MITR FUEL BY GAMMA-RAY SPECTROSCOPY. Massachu-
     setts Institute of Technology. Scientific Report no. 4.
     Oct. 1965. 261p. Thesis
                              Also report AFCRL-65-787; MITNE-64

696. Srnka, D. PREPARATION AND USE OF LITHIUM-DRIFTED GERMANIUM DE-
     TECTORS. Ceskoslovenska Akademie Ved. Rez. Ustav Jaderneho
     Vyzkumu. Aug. 1967. 23p.   UJV-1873

     Detailed fabrication techniques for planar and coaxial
     lithium-drifted germanium detectors are described.

697. Stab, L.; Henck, R.; Siffert, P., and Coche, A. PREPARATION OF
     N-I-P DIODES FROM GERMANIUM. APPLICATION TO GAMMA SPECTROM-
     ETRY. Nucl. Instrum. Methods, 35 (1965), 113-19.
     (In French)

     A method for the preparation of germanium n-i-p junctions
     is described, and results obtained in the use of such junctions
     in gamma spectrometry are discussed.

698. Staeudner, R. OPERATIONAL PRINCIPLES AND PHYSICAL PROPERTIES OF
     SEMICONDUCTOR DETECTORS. (WIRKUNGWEISE UND PHYSILALISCHE
     EIGEN SCHAFTEN VON HALBLEITERDETEKTOREN). In Soc. Europeene
     de Protec. Contre des Rayonnements Detectors in Radiation
     Protec. and Radiation Measurement Technology. Sept. 1967.
     7-16. (In German)

699. Steffen, R. M. RESEARCH IN NUCLEAR PHYSICS. Purdue University,
     Lafayette, Ind., Dept. of Physics. Final Report, Oct. 1,
     1966-Sept. 30, 1967. March 29, 1968. 90p.   COO-1420-139

     Some developments in instrumentation, including methods for
     fabricating Ge(Li) detectors, are summarized.

700. Strauss, M. G.; Sherman, I. S.; Brenner, R.; Rudnick, S. J.;
     Larsen, R. N., and Mann, H. M. HIGH RESOLUTION GE(LI)
     SPECTROMETER FOR HIGH INPUT RATES. Rev. Sci. Instrum., 38,
     no. 6 (June 1967), 725-730.

     A system was developed for obtaining gamma-ray spectra with
     high resolution at high input rates. For rates up to 100,000
     gamma-ray events/sec in a 3.5 cm^3(13 mm thick) Ge(Li) detector,
     the FWHM at 1.33 MeV is 2-3$^{1/2}$ keV and the spectral shift is no

more than 0.1%. The salient features of the system are described
and the performance data are presented and discussed.

701. Strauss, M. G. and Larsen, R. N. PULSE HEIGHT DEFECT DUE TO
     ELECTRON INTERACTION IN THE DEAD LAYERS OF GE(LI) GAMMA-RAY
     DETECTORS. Nucl. Instrum. Methods, 56 (1967), 80-92.

702. Strauss, M. G.; Larsen, R. N., and Sifter, L. L. PULSE SHAPE
     DISTRIBUTIONS FROM GAMMA-RAYS IN LITHIUM DRIFTED GERMANIUM
     DETECTORS. Nucl. Instrum. Methods, 45 (1957), 45-54.

     Pulse shape distributions resulting from the interaction of
     gamma-rays in planar germanium detectors were studied as a func-
     tion of field intensity, gamma-ray energy, and detector width.

703. Strauss, M. G.; Lenkszus, F. R., and Eichholz, J. J. SIMPLE AND
     ACCURATE CALIBRATION TECHNIQUE FOR MEASURING GAMMA-RAY ENER-
     GIES AND GE(LI) DETECTOR LINEARITY. Nucl. Instrum. Methods,
     76 (1969), 285-94.

     A technique for precise calibration of Ge(Li) gamma-ray spec-
     trometers which is particularly useful in studies of decay
     schemes, has been developed.

704. Strauss, M. G.; Sifter, L. L.; Lenkszus, F. R., and Brenner, R.
     ULTRA STABLE REFERENCE PULSER FOR HIGH RESOLUTION SPECTROM-
     ETERS. IEEE Trans. Nucl. Sci., NS-15, no. 3 (June 1968),
     518-530.

705. Suominen, Pekka. DESIGN, CONSTRUCTION, AND CHARACTERISTICS OF
     SOME GE(LI) DETECTORS AND SPECTROMETERS. Jyvaskyla Univ.,
     Finland. Dept. of Physics. Aug. 1970. 115p. Thesis
                                            Also report NP-18653

     Experience obtained in preparing different kinds of Ge(Li)
     detectors is presented. The preparation method used involves
     an unusual sodium hypochlorite etching procedure. A method for
     externally cooling the field effect transistor of a standard
     preamplifier is presented and some test results are described.
     A "semitoroidal" detector configuration intended for making large
     Ge(Li) detectors with approximately homogeneous electric fields
     is described in detail. Test measurements have been carried out
     with a relatively small semitoroidal detector. Some data ob-
     tained with an anti-Compton spectrometer, which results in a

peak-to-tail ratio of 200 for ^{137}Cs spectrum, are presented. Cryogenic and electronic problems concerning the performance of the anti-Compton spectrometer are discussed in detail. Some two-crystal pair spectrometers have been tested and possibilities for further development are discussed. Efficiency and line shape problems involved in a summing Compton spectrometer have been examined both theoretically and experimentally. On the basis of optimizing work a summing Compton spectrometer has been constructed. A very simple "analog on-line-computer" has been employed in connection with the spectrometer which renders it possible that gamma quanta with energies down to 100 keV might be recorded. The spectrometer resulted in spectra with higher peak-to-tail ratios than the ratios obtained with anti-Compton spectrometers. Further development of the summing Compton spectrometer employing, for instance, a Si(Li)-Ge(Li) detector pair is discussed.

706. Suominen, P. and Kantele, J. SOME HIGH-EFFICIENCY TWO-CRYSTAL GE(LI)-NAI(TL) PAIR SPECTROMETERS. Nucl. Instrum. Methods, 58 (Jan. 1968), 229-35.

Two types of two crystal pair spectrometers employing Ge(Li) principle and NaI(Tl) auxiliary detectors are described.

707. Taff, L. M. and Champion, P. M. COMPUTER PROGRAM FOR AUTOMATIC ANALYSIS OF SEMICONDUCTOR DETECTOR SPECTRA. Ames Laboratory, Iowa. Nov. 1968. 66p.   IS-1986

A computer program is described which accepts a Ge(Li) detector spectrum, finds peaks in the presence of statistical fluctuations in the data using the method of Mariscotti, fits each peak a Gaussian and quadratic background function, and produces convenient tabular results and plotted graphs. Ease of use and adaptability were major programming considerations. Complete instructions and a listing of the program are given.

708. Takacs, J. THE DEPLETION DEPTHS OF LITHIUM-ION DRIFT DETECTORS AS A FUNCTION OF THE TIME, VOLTAGE AND THE DIFFUSION COEFFICIENT. Nucl. Instrum. Methods, 33 (Mar. 1, 1965), 174-75.

The depletion depth (w) of lithium-ion drift detectors was calculated from the equation $W = (2_u Vt/kT)$, where u is the mobility, V is the drifting voltage, and t is the time. The mobility and the diffusion coefficient D are connected by Einstein's relation $D = (kT/e)\mu$. The results of the calculations are presented in the form of nomographs from which the depletion depths

as a function of time can be read for different drift voltage
and temperature.

709. Tambovtsev, D. I. and Kozlovskii, L. K. TESTS OF SURFACE-BARRIER
     SEMICONDUCTOR DETECTORS AT LIQUID HELIUM TEMPERATURES. Pri-
     bory Tekh. Eksp., no. 5 (Sept.-Oct. 1969), 59-61. (In
     Russian)

     It was shown that surface-barrier Si and Ge detectors of
standard fabrication were completely reliable down to T=4.2°K.
Only thermostable contacts were necessary. Several construction
modifications were proposed for the detectors.

710. Tamm, U.; Michaelis, W., and Coussieu, P. A PULSE SHAPE DISCRIM-
     INATION CIRCUIT FOR LITHIUM-DRIFTED GERMANIUM DIODES. Nucl.
     Instrum. Methods, 48 (1967), 301-305.

     A pulse-shape discrimination circuit has been developed which
is sensitive to a slow time-constant component in the charge
carrier collection of lithium-drifted germanium detectors.

711. Tavendale, A. J. CHARACTERISTICS OF SOME LARGE, COAXIAL LITHIUM-
     DRIFT SEMICONDUCTOR GAMMA-RAY SPECTROMETERS. IEEE Trans.
     Nucl. Sci., NS-13, no. 3 (June, 1966), 315-27.

     Large-volume, Li-drifted Ge and Si diodes were fabricated
using the coaxial configuration and their characteristics as
diodes and gamma spectrometers were investigated.

712. Tavendale, A. J. COAXIAL LITHIUM-DRIFTED GERMANIUM DIODES FOR
     GAMMA-RAY SPECTROSCOPY: FABRICATIONS AND CHARACTERISTICS.
     pp. 4-28 of "Panel on Lithium-Drifted Germanium Detectors."
     Vienna, International Atomic Energy Agency, 1966.

713. Tavendale, A. J. A GERMANIUM LITHIUM-DRIFT DIODE WITH GUARD-RING
     FOR GAMMA-RAY SPECTROSCOPY. Nucl. Instrum. Methods, 36 (Oct.
     1965), 325-27.

     The fabrication of deep-depletion-layer (greater than about
7 mm) Li-drifted Ge p-i-n diodes is described. The starting
material is p-type 10-ohm cm Ga-doped Ge. These diodes are used
for gamma spectroscopy. A guard ring technique is used to reduce to surface leakage current below 0.1 m A.

714. Tavendale, A. J. and Ewan, G. T. A HIGH-RESOLUTION LITHIUM-
     DRIFT GERMANIUM GAMMA-RAY SPECTROMETER. Nucl. Instrum.
     Methods, 25 (1963), 185-87.

     A discussion of the properties of the detectors fabricated
     and used as a gamma ray spectrometer by the Chalk River Atomic
     Energy Nuclear Lab.

715. Tavendale, A. J. A LARGE COAXIAL LITHIUM-DRIFT GERMANIUM DIODE
     FOR GAMMA-RAY SPECTROSCOPY. Australian Atomic Energy Com-
     mission Research Establishment, Lucas Heights. Sept. 1965.
     14p.   AETC/TM-299; CONF-650839

     From the Conference on Nuclear Physics, Melbourne.
     A "single-ended", coaxial, lithium-drift germanium diode with
     an active volume ~30 cm^3 (1.8 in.^{3}) was fabricated and tested as
     a high resolution gamma-ray spectrometer.

716. Tavendale, A. J. LARGE GERMANIUM LITHIUM-DRIFT P-I-N DIODES FOR
     GAMMA RAY SPECTROSCOPY. IEEE Trans. Nucl. Sci., NS-12, no.
     1 (1965), 255.

     Germanium lithium ion drift p-i-n diodes with depletion
     depths up to 11 mm and active volume 6 cm^3 have been fabricated
     from 10 ohm-cm, gallum doped material using a high-power drift
     method.

717. Tavendale, A. J. LARGE VOLUME LITHIUM-DRIFT DIODES. Canadian
     Patents, 759, 633. May 23, 1967.

     Large volume lithium-drift diode detectors may be produced
     as follows. Lithium is diffused onto at least two surfaces of
     a crystal of p-type material. One surface and the two opposing
     surfaces do not receive a coating. A bias voltage is applied
     between the lithium coated surfaces and a contact placed cen-
     trally on an end surface not diffused with lithium. In a
     controlled atmosphere, heat is applied such that drifting takes
     place inwardly from the lithium-diffused surface leaving a co-
     axial core of p-type material of small size in the central re-
     gion.

718. Tavendale, A. J. LOW-NOISE, CHARGE-SENSITIVE, VACUUM TUBE
     PREAMPLIFIER FOR SEMICONDUCTOR DETECTOR. Atomic Energy of
     Canada Ltd., Chalk River, Ontario. Sept. 1964. 11p.
                                      AECL-2071; CRGP-1182

719. Tavendale, A. J.   METHOD AND APPARATUS FOR PRODUCTION OF SEMI-
        CONDUCTOR LITHIUM-ION DRIFT DIODES.   Feb. 7, 1967.   Canadian
        Pat. 752, 583.

720. Tavendale, A. J. and Fowler, I. L.   PROGRESS ON A FAST-DRIFT METHOD
        FOR MAKING LITHIUM-ION-DRIFT GERMANIUM P-I-N GAMMA RAY SPEC-
        TROMETERS.   Chalk River Nuclear Laboratory.   Nov. 1964.   10p.
                                                        AECL-2110; GPI-57

        A fast-boiling liquid method used to drift large-volume
    (6 cm^3), deep-depletion-layer (up to 11 mm), p-i-n germanium di-
    odes for use as high-resolution gamma spectrometers is described.
    The method is about five times as fast as that obtained in an
    air-drift apparatus.

721. Tavendale, A. J.   REDUCTION OF CARRIER TRAPPING IN GERMANIUM
        LITHIUM-DRIFT GAMMA RAY SPECTROMETERS.   Rev. Sci. Instrum.,
        36 (Aug. 1965), 1275-6.

        It was previously found that charge-carrier trapping led to
    low-energy tails on the full energy peaks gamma spectra, impair-
    ing the energy resolution.   An investigation of pairs of com-
    parable lithium-drifted germanium diodes showed that a post-
    drift high temperature cycle the carrier trapping.

722. Tavendale, A. J.   RESTORATION OF LITHIUM DRIFTABILITY IN SOME
        VACUUM-GROWN GERMANIUM CRYSTALS FOR GAMMA-RAY DETECTORS.
        IEEE Trans. Nucl. Sci., NS-17, no. 3 (June 1970), 130-8.
                                                    Also report CONF-700301

723. Tavendale, A. J.   SEMICONDUCTOR LITHIUM-ION DRIFT DIODES AS HIGH-
        RESOLUTION GAMMA-RAY PAIR SPECTROMETERS.   IEEE Trans. Nucl.
        Sci., NS-11 (1964), 191-200.

        Germanium and silicon p-i-n diodes with active volumes up to
    2 cm^3 have been fabricated using the lithium-ion drift technique,
    p-type germanium doped with either gallium or zinc (5-10 ohm cm)
    and baron-doped silicon (1100 ohm cm) were the base materials.

724. Tavendale, A. J.   SEMICONDUCTOR NUCLEAR RADIATION DETECTORS. Aus-
        tralian Atomic Energy Commission Research Est., Lucas
        Heights.   Annu. Rev. Nucl. Sci., 17 (1967), 73-96.

725. Tavendale, A. J. TIMING RESPONSE VARIATION WITH TEMPERATURE FROM
        A PLANAR GERMANIUM (LITHIUM) DETECTOR. pp. 130-7 of "Semi-
        conductor Nuclear-Particle Detectors and Circuits." Brown,
        W. L., ed. Washington, D. C., National Academy of Sciences,
        1969.   Also report CONF-670520

        From the Conference on Semiconductor Nuclear Particle Detec-
        tors and Circuits, Gatlinburg, Tenn.
        Timing responses were measured for a planar contact Ge(Li)
        detector at 124, 78, and 30°K. The measurements indicate that
        the time response improves with decreasing temperature.

726. Taylor, James Murray. SEMICONDUCTOR PARTICLE DETECTORS. Wash-
        ington, D. C., Butterworth, 1963. 175p.

        A volume devoted to the design and performance of semicon-
        ductor detectors, giving their advantages and their limitations.

727. Terada, Jitsuo and Takayanagi, Seiichi. LITHIUM-DRIFTED GERMAN-
        IUM DETECTORS. Toshiba Rebyu, 21 (Feb. 1966), 176-82.
        (In Japanese)

        The encapsulated lithium-drifted germanium detectors have
        been developed with gamma resolution of 3.2 keV at 122 keV. A
        simple cryostat assembly for the encapsulated detectors has also
        been studied.

728. Tharun, U. and Kaffrell, N. DIGITALE FENSTER FUER GAMMA-GAMMA-
        KOINZIDENZMESSUNGEN MIT HALBLEITERDETEKTOREN. (DIGITAL WIN-
        DOW FOR GAMMA-GAMMA COINCIDENCE MEASUREMENTS WITH SEMICON-
        DUCTOR DETECTORS.) Nucl. Instrum. Methods, 62 (1968), 184-8.

        Digital windows have been designed and built to avoid the
        disadvantages of normal single channel analysers setting the gate
        in gamma-gamma-coincidence experiments with two Ge(Li) detectors.

729. Thomas, Jess Brooks, Jr. MEASUREMENT OF NUCLEAR LIFETIMES BY THE
        DOPPLER ATTENUATION SHIFT METHOD. Stanford University. 1968.
        84p. Thesis

        The Doppler shift attenuation method, including centroid and
        shape analysis, is developed in detail. The lifetimes of the
        Li^7 (478), Be^7 (432), shift attenuation method and lithium-
        drifted germanium detectors. The associated radiative widths
        are compared in each case with the widths predicted by shell
        model theory.

730. Thompson, A. C. and Dalby, D. A. A DRIFT CURRENT CONTROL FOR
     GERMANIUM DETECTOR PREPARATIONS. Nucl. Instrum. Methods,
     51 (1967), 178-180.

     A simple inexpensive control circuit and drift unit for
     constant-power-drift planar detectors are described. A uni-
     junction transistor oscillator controls an SCR and offers ad-
     justable maximum temperature and sensitivity.

731. Thompson, K. M. and Gruhn, C. R. HIGH PRECISION GONIOMETER FOR
     CHARGE PARTICLE SPECTROSCOPY. Nucl. Instrum. Methods, 74
     (1969), 309-14.

     A high precision goniometer capable of positioning Ge(Li)
     charged particle detectors with a precision of 0.02° has been
     built. This device consists of a 4 ft radius, detector support
     which rotates outside of a central target vacuum chamber. The
     main features of the system are described and a summary of the
     tests to verify its operational capabilities is presented. A
     description is also given of the installation of the unit on the
     beam line, illustrating the precautions which were taken to en-
     sure alignment. The performance of this apparatus in studying
     Ni(p,p) reactions at 40 MeV is discussed.

732. Tokcan, G. and Cothern, C. R. EFFICIENCY DETERMINATIONS FOR GE
     (LI) DETECTOR. Ohio J. Sci., 69 (Mar. 1969), 92-100.

733. Tokcan, G. and Cothern, C. R. PRECISE FORM OF THE EFFICIENCY
     EQUATION FOR A GE(LI) DETECTOR. Nucl. Instrum. Methods, 64
     (1968), 219-220.

     The applicability of the semi-empirical relation used by
     Freeman and Jenkins to lower energies (down to 200 keV) is
     examined.

734. Tolmie, R. W. and Thompson, C. J. FIELD EQUIPMENT FOR NEUTRON
     ACTIVATION ANALYSIS. pp. 489-505 of "Nuclear Techniques and
     Mineral Resources." Vienna, International Atomic Energy
     Agency, 1969.   Also report STI/PUB-198; CONF-681117

735. Trammell, R. and Walter, F. J. COMPARISONS OF DETECTOR OUTPUT
     PULSE HEIGHTS. pp. 221-6 of "Semiconductor Nuclear-Particle
     Detectors and Circuits." Brown, W. L., ed. Washington,
     D. C., National Academy of Sciences, 1969.
                                   Also report CONF-670520

From the Conference on Semiconductor Nuclear Particle Detector and Circuits, Gatlinburg, Tenn.

In placing Ge(Li) detectors in parallel in close-packed
arrays the problem of variations in detector charge output from
unit to unit is encountered. In order to obtain a preliminary
insight into the problem of obtaining closely matched detectors,
the output pulse heights for 33 consecutively produced Ge(Li)
detectors were measured. A significant spread in pulse heights
was found.

736. Trammell, R. and Walter, F. J. THE EFFECTS OF CARRIER TRAPPING
IN SEMICONDUCTOR GAMMA-RAY SPECTROMETERS. Nucl. Instrum.
Methods, 76 (1969), 317-21.

737. Turcotte, R. E. and Moore, R. B. SIMPLE VERSATILE CRYOSTAT FOR
ENCAPSULATED DETECTORS. Nucl. Instrum. Method, 72 (1969),
210-12.

A simple, inexpensive and versatile cryostat for encapsulated
detectors is described. The cryostat operates at liquid nitrogen temperatures with molecular sieve pumping. Standard performance was: pressure less than 5×50^{-5} mm, temperature of
detector less than 80°K, consumption of liquid nitrogen about
1 liter per day.

738. Tuzzolino, A. J. OUTPUT PULSE AMPLITUDE FROM LITHIUM-DRIFTED
DETECTOR-AMPLIFIER COMBINATION UNDER CONDITIONS OF CARRIER
DIFFUSION AND BULK AND SURFACE RECOMBINATION. J. Appl.
Phys., 42, no. 3 (March 1971), 1185-91.

A calculation has been made of the output pulse amplitude
obtained from an idealized lithium-drifted detector-amplifier
combination under conditions where an incident charge particle
generates a charge distribution in the field-free region of the
detector which is a general function of position.

739. Tyree, William H. and Bistline, Robert W. ^{239}Pu DETECTION IN
VIVO WITH A GERMANIUM LITHIUM ARRAY. Dow Chemical Co.,
Golden, Colo. June 12, 1970. 8p.   REP-1488

Results are given of simulated in vivo measurements of plu-
tonium-239 amounts detected in the chest (lung) cavities of a
water filled phantom model. The device developed is a germanium-
lithium detector array, using the 51.63 kiloelectron-volt gamma
ray from plutonium-239.

740. Uehara, Shinichi; Kawase, Yoichi, and Okano, Kotoyuki. EFFI-
      CIENCY CURVES OF GE(LI) DETECTORS AND AN INTERNAL CONVERSION
      COEFFICIENT SPECTROMETER. Kyoto University (Japan). Re-
      search Reaction Institute. Dec. 20, 1969. 15p. (In Japa-
      nese)   KURRI-TR-71

741. Ungrin, J. and Johns, M. W. GERMANIUM X-RAY ESCAPE PEAKS IN THE
      40 to 411 KEV RANGE PRODUCED BY SMALL WINDOWLESS GE(LI) DE-
      TECTORS. Nucl. Instrum. Methods, 70 (Apr. 1, 1969), 112-14.

      Germanium x-ray escape peaks have been observed for gamma rays
      in the energy range 40 to 411 keV chg using a small windowless
      Ge(Li) detector 3 mm thick. The ratio of the area of the escape
      peak to that of the photo-peak is ~0.24% and almost independent
      of gamma ray energy for energies above ~150 keV.

742. Van Assche, P. H. M. HIGH RESOLUTION MEASUREMENTS OF GAMMA RAYS
      FROM (n, γ) REACTIONS. Rev. Phys. Appl., 4 (June 1969),
      269-70.   Also report CONF-681208-Suppl.

      A comparison has been made between a semiconductor detector
      and a diffraction spectrometer when used in the analysis of neu-
      tron capture gamma radiation.

743. Van Der Leun, C. and DeWit, P. A SIMPLE AND VERSATILE GAMMA-RAY
      SOURCE FOR THE CALIBRATION OF HIGH ENERGY GE(LI) SPECTRA.
      Phys. Lett., 30B (Nov. 10, 1969), 406-8.

      A simple and inexpensive neutron-capture source of nonoener-
      getic high-energy gamma-rays of many different energies suited
      for the energy-calibration of Ge(Li) spectra, is described. An
      americium-241-beryllum neutron source is used.

744. Van Patter, D. M. A COINCIDENCE-ANTICOINCIDENCE GE(LI)-NAI(TL)
      SYSTEM. pp. 99-108 of "Proceedings of the Symposium on Nu-
      clear Physics Research with Low Energy Accelerators." New
      York, Academic, 1967.

745. Varnell, L. and Trischuk, J. A PEAK-FITTING AND CALIBRATION PRO-
      GRAM FOR GE(LI) DETECTORS. Nucl. Instrum. Methods, 76 (1969),
      109-14.

      A program is described for the analysis of gamma-ray spectra
      obtained with Ge(Li) spectrometers. There are several advantages

to this program (1) the function used gives a better fit to the
experimental peak shape, and (2) unknown sources are fitted
with the same function and the actual peak shape are minimized.
Using this program, relative intensities have been measured to
better than one percent, and energies better than 0.1 keV.

746. Vervier, J. THE USE OF GE(LI) DETECTORS IN NUCLEAR PHYSICS.
     Paper 4 of "Proceedings of the Meeting on Applications of
     Ge(Li) Detectors in Science, Technology, Medicine and
     Industry, Brussels." Oct. 1967. 23p.
                                        BLG-425; CONF-671078

747. Voight, Adolf F.; Becknell, Duane E., and Menapace, Laurene.
     COMPARISON OF SOLID STATE AND SCINTILLATION GAMMA-RAY SPEC-
     TROMETRY IN ANALYSIS. pp. 1035-42 of the "Conference on
     Modern Trends in Activation Analysis." vol. 2. DeVoe, J.
     R., ed. Washington, D. C., National Bureau of Standards,
     1969.   Also report CONF-681003-(vol. 2)

     The Ge(Li) and NaI(Tl) scintillation detectors are compared
     as to precision in the determination of several elements in the
     nonstoichiometric W "bronzes" and the analysis of mixtures of
     rare earths. Results for the determination of K in K-W bronzes
     and for La in La-W bronzes are tabulated. Binary mixtures of
     rare earths were analyzed with good precision using the Ge(Li)
     detectors, but the spectral resolution was not good with a
     ternary earth mixture. No definite conclusions are drawn as to
     which detector is best, since both seem to have merit in dif-
     ferent situations.

748. Voinova, N.; Vrzal, Ya.; Dzhelepov, B.; Liptak, Ya., and Urbanets,
     Ya. INVESTIGATION OF ^{182}Ta GAMMA-IRRADIATION BY USING A
     SEMICONDUCTOR SPECTROMETER. (ISSLEDOVANIE GAMMA-IZLUCHENIYA
     ^{182}Ta NA POLUPROVODNIKOVOM SPEKTROMETRIE.) Jt. Inst. for
     Nuclear Research. Dubnar (USSR) Laboratory of Neutron Physics.
     Oct. 1966. 37p.   JINR-P6-2976

     The gamma spectrum of ^{182}Ta for the energy range 50-1500 keV
     was investigated using a semiconductor spectrometer with a Ge(Li)
     detector.

749. Vrzal, Ya.; Dzhelepov, B. S.; Liptak, Ya.; Moskvin, L. N., and
     Prikhodtseva, V. P. GAMMA-RADIATION OF ^{69}GE (GAMMA-IZLUCH-
     ENIE ^{69}GE). Joint Inst. for Nuclear Research, Dubna (USSR).
     Lab. of Theoretical Physics. 1968. 12p. (In Russian)
                                        JINR-P6-3698

750. Wachsmann, F. and Drexler, G. MEASUREMENT OF THE X-RAY SPECTRA
     BY MEANS OF GERMANIUM SEMICONDUCTOR DETECTORS. Magy. Radiol.,
     20 (June 1968), 194-201. (In Hungarian)

     The possibilities of the use of germanium semiconductor
     detectors, drifted with lithium, in the energy-range of 10-150
     keV are outlined. The spectra obtained by means of such detec-
     tors, were easily smoothed and the method was very simple with
     the appropriate apparatus. The spectra of primary radiations
     filtered in various degree are discussed. The changes of the
     spectrum of the x-ray passed through layers of various thickness,
     whose substance were identical with that of the body, was demon-
     strated.

751. Wagner, Ronald S.; Tyler, III, Mark V., and Lee, Yung-Keun.
     LABORATORY FABRICATION OF LITHIUM-DRIFTED GERMANIUM DETECTORS
     FOR GAMMA RAYS. pp. 150-158 of "Johns Hopkins University,
     Dept. of Physics. Nuclear Reactions in Nuclei and Moess-
     bauer Studies." Sept. 1967.   NYO-2028-2

     Production of Ge(Li) detectors which meet the individual re-
     quirements of several projects in our laboratory was continued.
     In particular, the thin-window Ge(Li) detectors for soft gamma-
     rays were developed for Moessbauer experiments. Progress has
     been made in fabricating large area thin-window detectors and
     cooled-FET preamplifiers.

752. Wahlgren, M. A.; Strauss, M. G., and Hines, J. J. HIGH-EFFI-
     CIENCY GAMMA-RAY SPECTROMETER USING TWO SEPARATE COAXIAL
     GE(LI) DETECTORS. Trans. Amer. Nucl. Soc., 12 (June 1969),
     67.   Also report CONF-690609

753. Wainio, K. M. and Knoll, G. F. CALCULATED GAMMA RAY RESPONSE
     CHARACTERISTICS OF SEMICONDUCTOR DETECTORS. Nucl. Instrum.
     Methods, 44 (1966), 213-214.

     A Monte Carlo computer program has been used to calculate
     characteristics of the response of fully depleted silicon and
     germanium radiation detectors to monoenergetic gamma rays. Data
     for total absorption probability, intrinsic efficiency, escape
     peak efficiency and pulse height spectra are presented as func-
     tions of detector thickness and photon energy. Other parameters
     of interest in analysing detector response are also given. The
     results of a second Monte Carlo calculation of electron migra-
     tion in silicon and germanium are employed to account for the
     leakage of secondary electrons from the detector volume.

Bremsstrahlung energy loss by electrons is also simulated. The calculations are expected to be applicable in those cases in which secondary electron energies do not exceed 2 MeV. Comparison with experiment shows good agreement within this limitation.

754. Walford, G. and Doust, C. E. ANOMALOUS EFFECT IN A LITHIUM-DRIFTED GERMANIUM SEMICONDUCTOR. Electron. Lett., 4 (1968), 13.

The characteristics of a lithium-drifted germanium gamma-ray detector in a vacuum cryostat at 77°K are described.

755. Walford, G. and Doust, C. E. DEFECTS IN GE(LI) DETECTOR EFFICIENCIES. pp. 103-18 of the "Proceedings of the Meeting on Special Techniques and Materials for Semiconductor Detectors, Ispra, Italy, 1968." June 1969.   EUR-4269

A study was made of the variation of active volume and full peak efficiency of Ge(Li) detectors. It was found that only high resolution detectors had the expected high efficiencies. An eddy current technique is proposed for investigating the internal variation of the p-i-n contours.

756. Walford, G. A DETERMINATION OF THE RELATIONSHIP BETWEEN EFFICIENCY AND VOLUME OF A GE(LI) DETECTOR AND A PROPOSED EDDY CURRENT TECHNIQUE FOR DEPTH TESTING THE P-I-N CONTOURS OF THE CRYSTAL. Nucl. Instrum. Methods, 67 (Jan. 15, 1969), 272-6.

A study was made of the variation of active volume and full peak efficiency of Ge(Li) detectors. It was found that only high resolution detectors had the expected high efficiencies. Also an eddy current technique is proposed for investigating the internal variation of the p-i-n contours.

757. Walford, G. and Doust, C. E. A METHOD FOR RAPID CALIBRATION OF GERMANIUM SPECTROMETERS. Nucl. Instrum. Methods, 62 (1968), 353-354.

Discusses the energy and total peak efficiency calibration of germanium spectrometers using radium with its decay products.

758. Walker, David-Marshall. AN INVESTIGATION OF MULTIPLE GAMMA
     SCATTERING IN GERMANIUM AS APPLIED TO GE(LI) GAMMA SPECTROM-
     ETERS. Atlanta, Georgia Inst. of Technology. 1970. 175p.
     Thesis

759. Walker, D. M. and Palms, J. M. MONTE-CARLO ANALYSIS OF THE GE(LI)
     DETECTOR USED IN THE SUM-COINCIDENCE MODE. IEEE Trans.
     Nucl. Sci., NS-17, no. 3 (June 1970), 296-305.
                                        Also report CONF-700301

760. Walker, W. W.; Moak, C. D., and Dabbs, J. W. T. RESPONSE OF GER-
     MANIUM AND SILICON DETECTORS TO ENERGETIC BROMINE AND IODINE
     IONS. pp. 12-127 of "Oak Ridge National Laboratory, Tenn.
     Physics Division Annual Progress Report Period Ending Dec.
     31, 1964." May 1965.

     A number of germanium surface-barrier detectors were fabri-
     cated and exposed to energetic [79]Br and [127]I ions from a Van de
     Graeff accelerator, to spontaneous fission fragments from [252]Cf
     and to alpha particles from [241]Am and [252]Cf. Seven of ten de-
     tectors gave good resolutions of about 20 keV for alpha parti-
     cles at 77°K. For purposes of comparison, several silicon de-
     tectors were also studied along with the germanium detectors.

761. Wallace, G. and Coote, G. E. EFFICIENCY CALIBRATION OF GE(LI)
     DETECTORS USING A RADIUM SOURCE. Nucl. Instrum. Methods,
     74 (1969), 353-4.

     The decay of radium and its daughters has been re-investi-
     gated for use in relative efficiency calibration of Ge(Li) de-
     tectors.

762. Walter, F. J. INTRODUCTORY REMARKS ON GAMMA DETECTION EFFI-
     CIENCY OF GERMANIUM SPECTROMETERS. pp. 214-20 of "Semi-
     conductor Nuclear-Particle Detectors and Circuits." Brown,
     W. L., ed. Washington, D. C., National Academy of Sciences,
     1969.   Also report CONF-670520

     The efficiencies of seven consecutively produced cylindrical
     coaxial Ge(Li) detectors were measured and the values ranged
     from 2.4 to 2.7%. The problems of constructing a definition
     and geometry independent measuring technique for detector effi-
     ciencies is discussed. The problems of geometry factors, sur-
     face-to-volume ratio, and mysterious volume losses are dis-
     cussed.

763. Waugh, John B. S. LEADING EDGE TIMING CIRCUIT FOR GE(LI) DETEC-
        TOR. IEEE Trans. Nucl. Sci., NS-15, no. 3 (June 1968),
        509-517.

        A circuit for producing a timing signal from the leading
     edge of slow rising Ge(Li) detector output signals is described.

764. Webb, P. P. EFFICIENCY CONSIDERATIONS IN GERMANIUM (LITHIUM)
        DETECTORS. pp. 227-9 of "Semiconductor Nuclear-Particle
        Detectors and Circuits." Brown, W. L., ed. Washington,
        D. C., National Academy of Sciences, 1969.
                                            Also report CONF-670520

765. Webb, P. P. and Green, R. M. ENCAPSULATED GERMANIUM GAMMA-RAY
        SPECTROMETERS: PERFORMANCE CHARACTERISTICS AND OPERATIONAL
        EXPERIENCE. IEEE Trans. Nucl. Sci., NS-13, no. 3 (June 1966),
        445-456.

766. Webb, P. P. and Williams, R. L. GAMMA-RAY SPECTROSCOPY USING A
        GERMANIUM LITHIUM DIODE. Nucl. Instrum. Methods, 22 (1963),
        361-2.

        Ge diodes with Li-drifted depletion layers are shown to be
     advantages for gamma ray spectroscopy. The 122 and 136 keV lines
     of the ^{57}Co spectrum are readily resolved from the Compton back-
     ground. The ^{137}Cs spectrum has a photo electric peak twice the
     height of the Compton background. It is pointed out that a Ge
     Li-drifted diode will require additional drifting after each use.

767. Webb, P. P. RELAXATION PHENOMENA IN PLANAR GE(LI) DETECTORS.
        IEEE Trans. Nucl. Sci., NS-15, no. 3 (June 1968), 321-326.

768. Webb, P. P. TRAPPING MEASUREMENTS IN GERMANIUM (LITHIUM) DETEC-
        TORS. pp. 138-43 of "Semiconductor Nuclear-Particle Detec-
        tors and Circuits." Brown, W. L., ed. Washington, D. C.,
        National Academy of Sciences, 1969.
                                            Also report CONF-670520

        Trapping measurements were performed for Ge(Li) detectors made
     by four different manufacturers using a ^{137}Cs gamma source. The
     measurements indicate some brands show a preferential trapping of
     electrons while others show a preference for hole trapping.

769. Webb, P. P.; Malm, H. L.; Chartrand, M. G.; Green, R. M.;
     Sakai, E., and Fowler, I. L. USE OF COLLIMATED GAMMA-RAY
     BEAMS IN THE STUDY OF GE(LI) DETECTORS. Nucl. Instrum.
     Methods, 63 (1968), 125-135.

     Useful studies of Ge(Li) detector can be made with well-col-
     limated beams of gamma-rays. Choice of gamma-ray energy (0.3 to
     1 MeV) and design of collimators are discussed.

770. White, D. H. and Birkett, R. E. A GE(LI)-GE(LI)-NAI(TL) COINCI-
     DENCE SPECTROMETER SYSTEM FOR (n, γ) STUDIES. Nucl. Instrum.
     Methods, 73 (1969), 260-8.

     A spectrometer system has been constructed to study 0.06 -
     12 MeV neutron-capture gamma-rays produced at the thermal-beam
     facility of the Livermore Pool Type Reactor. The system con-
     sists of two spectrometers, designed to cover the low-energy
     (0.06-2.5 MeV) and high-energy system consists of a Ge(Li)
     detector flanked by two NaI(Tl) pair spectrometer. These sys-
     tems are operated independently, or disengaged to obtain adequate
     coincidence efficiency. An automated pulser is used for linear-
     ity calibration and stabilization. Coincidence data are pro-
     cessed by a 4096-channel three-parameter pulse-height analyzer,
     buffered onto magnetic tape, and analyzed off-line by digital
     computers.

771. White, D. H. NEUTRON-CAPTURE GAMMA-RAY STUDIES OF CALCIUM ISO-
     TOPES WITH A GE(LI)-GE(LI)-NAI(TL) COINCIDENCE SPECTROMETER
     SYSTEM. University of California, Livermore, Lawrence Radi-
     ation Laboratory. April 28, 1969. 28p.
                                        UCRL-71676; CONF-690802

     A three-parameter spectrometer system is described, consisting
     of two Ge(Li) NaI(Tl) gamma-ray spectrometers, one designed for
     high energy and one for low-energy neutron-capture gamma-rays.
     These systems may be operated in various modes, both singly and
     in coincidence. Data were taken with certain calcium isotopes,
     and the methods of the data analysis are given. A method is
     described to show how tentative transitions are fit to the pro-
     posed level scheme by a linear least-squares program. Results
     are presented for Ca-44.

772. White, D. H.; John, W.; Saunders, B. G., and Groves, D. J. A
     STUDY OF FISSION GAMMA RAYS USING A GE(LI) DETECTOR IN CON-
     JUNCTION WITH A BENT-CRYSTAL-SPECTROMETER IN CAUCHOIS GEOME-
     TRY. University of California, Livermore, Lawrence Radia-

tion Laboratory. July 1966. 7p.

UCRL-70000; CONF-660906-24

The high-resolution capability of the bent-quartz-crystal
spectrometer in Cauchois geometry was extended above 400 keV
by replacing the photographic emulsion with a Ge(Li) detector
and using multichannel pulse height analysis. By proper collimation at the focal circle, extended spectral bands of fission
gamma rays are recorded which are free of Compton edges and
other secondary processes.

773. Wichner, R.; Armantrout, G. A., and Brown, R. G. GE CRYSTAL
       GROWTH AND EVALUATION AS GE(LI) DETECTOR MATERIAL. IEEE
       Trans. Nucl. Sci., NS-17, no. 3 (June 1970), 160-4.
                                          Also report CONF-700301

A comprehensive program of germanium single crystal growth
and the evaluation of such material for use as Ge(Li) detectors
is described.

774. Wilburn, C. and Edwards, W. D. OXYGEN-FREE ALUMINUM DOPED GER-
       MANIUM FOR LITHIUM DRIFTING. Dec. 1966. 2p.   AD-656671

A description is given of a method of growing germanium
crystals which are effectively oxygen-free and suitable for the
fabrication of gamma ray spectrometers by the lithium drift process. Aluminum is used as both p-type dope and as a getter for
oxygen.

775. Williams, D.; Snelling, G. F., and Pickup, J. A GAMMA-SPECTRUM
       STABILIZER WITH COMPENSATION FOR THE EFFECTS OF DETECTOR
       TEMPERATURE VARIATION. Nucl. Instrum. Methods, 39 (Jan.
       1966), 141-9.

A compact system of spectrum stabilization is provided, for
use principally with single channel systems, and including compensation for detector temperature variations. Various methods
for reference provision are discussed and the development of an
appropriate type described.

776. Williams, G. H. and Morgan, I. L. A GE(LI) GAMMA SPECTROMETER
       FOR (n, n'γ) MEASUREMENTS. Nucl. Instrum. Methods, 45
       (1966), 313-318.

777. Williams, R. L. DRIFTABILITY OF GERMANIUM CRYSTALS. pp. 207-
     13 of "Semiconductor Nuclear-Particle Detectors and Cir-
     cuits." Brown, W. L., ed. Washington, D. C., National
     Academy of Sciences, 1969.   Also report CONF-670520

778. Williamson, C. F. and Alster, J. A MINIATURE COOLING SYSTEM FOR
     GE(LI) SOLID STATE DETECTORS. Nucl. Instrum. Methods, 46
     (1967), 341-343.

     A cooling system using a Joule-Thompson cryostat is described
     for the convenient use of a Ge(Li) solid-state detector in a
     vacuum chamber or in an anti-coincidence annulus.

779. Winiger, P.; Huber, O., and Halter, J. SEMICONDUCTOR MEASUREMENTS
     OF FALLOUT. Helv. Phys. Acta., 41 (Aug. 15, 1968), 645-9.
     (In German)

     Two gamma-ray spectra of recent fission products measured by
     a NaI well-type crystal and a Ge(Li) drifted diode are compared.
     The high energy resolution of the small semi-conductor compensate
     nearly its low detection efficiency and allows a much more reli-
     able analysis of complex spectra within the monitoring of radio-
     activity. An example of an analysis is given.

780. Winn, W. G. and Sarantites, D. G. DIRECTIONAL-CORRELATION ATTEN-
     UATION FACTORS FOR GE(LI) GAMMA-RAY DETECTORS. Nucl. Instrum.
     Methods, 66 (1968), 61-69.

     An experimental method for the determination of the angular
     correlation attenuation factors, due to finite solid angle, for
     Ge(Li) detectors is offered. The method is tested by comparing
     the measured factors for a 3" x 3" NaI(Tl) detector with availa-
     ble calculated attenuation factors. Results for two coaxial
     Ge(Li) detectors is with active volumes of 20 and 30 cm^3 are pre-
     sented as a function of distance and energy of the gamma-rays.

781. Winn, W. G. and Sarantites, D. G. DIRECTIONAL-CORRELATION ATTENU-
     ATION FACTORS FOR GE(LI) DETECTORS: A COMPARISON OF EXPERI-
     MENTAL AND CALCULATED VALUES. Nucl. Instrum. Methods, 82
     (1970), 230-6.

     Refinements in measuring technique are utilized to improve
     a previously described method for experimentally determining the
     directional-correlation attenuation factors of Ge(Li) γ-ray de-
     tectors. Correction factors for a five-sided true coaxial 29 cm^3

Ge(Li) detector and a five-sided trapezoidal 25 cm³ Ge(Li) detector are determined using this method and agree well with recently available computer calculations for these values. Various
sources of error inherent in the method are discussed in some detail.

782. Yamazaki, T. and Ewan, G. T. METHOD OF MEASURING NANOSECONDS
      ISOMERIC STATES PRODUCED IN (PARTICLE, XN) REACTIONS. Nucl.
      Instrum. Methods, 62 (1968), 101-104.

      A simple method for measuring the lifetimes of gamma-transi-
      tions from nanoseconds isomeric states is described. The method
      depends on the phase grouping of beam pulses in a cyclotron and
      the fast timing properties of Ge(Li) detectors.

783. Young, F. C.; Figuera, A. S., and Pfeufer, G. ABSOLUTE GAMMA-
      RAY EFFICIENCY OF A 50 CM³ GE(LI) DETECTOR. Nucl. Instrum.
      Methods, 92, no. 1 (March 1, 1971), 71-5.

784. Yule, Herbert P. COMPUTATION OF LITHIUM-DRIFTED GERMANIUM DETEC-
      TOR PEAK AREAS FOR ACTIVATION ANALYSIS AND GAMMA RAY SPEC-
      TROSCOPY. Anal. Chem., 40 (Aug. 1968), 1480-6.

      Results of studies of net full energy peak are computation
      methods for activation analysis and gamma ray spectrometry are
      reported. Using a computer routine to search for any and all
      peaks in the spectrum, peak boundary channels were located by
      studying the behaviour of a smoothed spectrum and smoothed first
      derivative spectrum, each formed from the original spectrum.
      Net peak areas were computed from that portion of the spectrum
      enclosed by the peak boundary channels, overcoming changes in
      peak shape due to resolution losses and to other causes. This
      method gave accurate results for activation analysis, decay
      curve resolution, and other peak intensity studies.

785. Zettlemoyer, A. C. SURFACE PROPERTIES OF GERMANIUM SEMICONDUCTORS.
      Lehigh University, Bethlehem, Pa. Aug. 1966. 4p.
                                          REPT-4720-C; AD-649919

786. Ziemba, Francis P. NUCLEAR PARTICLE SEMICONDUCTOR DETECTORS FOR
      AEROSPACE APPLICATIONS. Solid State Radiation Inc., Los
      Angeles. n. d. 19p.   CONF-660207-20

A review is presented on four types of semiconductor detectors, including description, manufacturing techniques, and applications for aerospace research. These detectors are surface-barrier, diffused junction, lithium-compensated silicon, and lithium compensated germanium.

787. Ziemba, Francis P. RESEARCH AND DEVELOPMENT OF FABRICATION OF PIN AND LID DETECTORS AND THE DRY RUN INTERFACE SYSTEM. Final report of the Solid State Radiation, Inc., Los Angeles, Calif. Aug. 31, 1963. 88p.   UCRL-13106

Two types of semiconductor detectors for fast transient high intensity gamma-rays were designed, fabricated, and evaluated. The first type is based on the lithium ion drift (LID) technique having a sensitive depth between 1 mm and 1 cm and active areas between 0.04 cm^2 and 1 cm^2. The second type is the PIN detector with a sensitive depth of 250 microns and an active area between 0.04 cm^2 and 1 cm^2. Dry run apparatus was designed for testing the static and dynamic characteristics of these detectors under field conditions. The properties of three different series of field tested detectors are tabulated.

788. Ziock, K. and Ritter, R. C. PROGRESS REPORT ON PARTICLE DETECTION AND PARTICLE DETECTORS. Virginia University, Charlottesville, Va. Nov. 30, 1963. 4p.

An investigation is conducted of a method for increasing the active volume of lithium-drifted germanium semiconductor detectors. This method involves a combination of lithium-drifting and special geometrical configurations.

789. Zullinger, H. R. and Aitken, D. W. CHARGE COLLECTION EFFICIENCIES FOR LITHIUM-DRIFTED SILICON AND GERMANIUM DETECTORS IN THE X-RAY ENERGY REGION. Stanford University, California. High Energy Physics Laboratory. Feb. 1968. 19p.
                                AD-677011; AFOSR-68-1676

Theoretical calculations showing the charge collection efficiencies of Ge(Li) and Si(Li) planar detectors as a function of trapping lengths and incident X-ray energies are presented.

790. Zullinger, H. R.; Middleman, L. M., and Aitken, D. W. LINEARITY AND RESOLUTION OF SEMICONDUCTOR RADIATION DETECTORS. *IEEE Trans. Nucl. Sci.*, NS-16, no. 1 (Feb. 1969), 47-61.

The energy linearity for Si(Li) and Ge(Li) has been measured and compared with theory. Good agreement was obtained for x-ray attenuation factors μmd<100. The region of the germanium absorption edge was investigated and found to exhibit larger nonlinearities than predicted by the theory. The discrepancy is believed to be due to inefficient charge collection near the detector window leading to excessive fluctuations in the trapped charge calculations of the line broadening due to trapping have been shown to agree qualitatively with experimental values for the silicon detector.

AUTHOR INDEX

(Numbers refer to entries rather than pages)

A

Abecasis, S. M.   98
Abul   Faiz Mohammed   8
Adam, G.   210
Adams, F.   1-4
Adda, L. P.   5
Ahmad, A. A. Z.   7, 9, 391
Ahmed, N. M.   7-9, 391
Aitken, D. W.   789, 790
Albridge, R. G.   638
Alexander, T. K.   10, 11, 110, 111, 464
Alkhozor, G. D.   12
Allen, B. J.   13-16, 86
Allen, S. J.   555
Allkofer, O. C.   17
Almen, O.   675
Alster, J.   778
Ammerlaan, C. A. J.   18, 19
Anders, O. U.   21
Andersen, B. V.   22, 23
Anderson, C. F. L.   304
Anderson, W. R.   24
Andersson, G. I.   236
Andrieux, C.   20
Anicin, I. V.   25
Antman, S. O. W.   26
Antonov, A. S.   27
Armantrout, G. A.   28-40, 129, 773
Arnell, S. E.   41, 675
Arsentev, I. N.   42
Asikainen, M.   43
Atomic Energy of Canada 44-47
Atzmony, U.   52
Aubin, G.   48, 49
Auble, R. L.   50
Auchampaugh, G. F.   51
Avida, R.   52
Avignone, F. T. III   53
Azuma, R. E.   445

B

Baedecker, P. A.   304
Baertsch, R. D.   54, 55, 331
Baicker, J. A.   665
Bailey, N. A.   56
Bakev, S. I.   674
Baldinger, E.   57-60
Baldwin, T. O.   61, 62
Balland, J.-C.   63-67
Barker, J. R.   68
Barker, P. H.   69
Barnes, B. K.   578
Baroody, A. J. Jr.   120
Barrette, J.   48, 49, 70
Barrette, M.   49
Bartholemew, G. A.   236
Battelle-Northwest, Richland, Wash.   71
Battleson, K.   440, 477
Baum, J. J.   72
Baumgaertner, M.   519
Beach, P. M.   335
Beck, R. N.   365
Beck, V. N.   569
Becknell, D. E.   747
Beery, D. B.   50
Beghian, L. E.   73
Bengtson, B.   539
Benoit, R.   74
Benson, K. E.   5
Beraud, R.   75
Berg, R. E.   76
Berkes, I.   75
Bernthal, F. M.   399
Berry, D. B.   50
Bertolini, G.   77-79
Bertrand, F. E.   80
Berzins, G.   50, 320
Beyer, L. M.   50
Bigham, C. B.   197
Bikit, I. S.   25

S

SUBJECT INDEX

(Numbers refer to entries rather than pages)